S0-AUG-152

PION-NUCLEON SCATTERING

series:

INVESTIGATIONS IN PHYSICS

Edited by ROBERT HOFSTADTER

1. Ferroelectricity by E. T. JAYNES
2. Mathematical Foundations of Quantum Mechanics by JOHN VON NEUMANN
3. Shell Theory of the Nucleus by EUGENE FEENBERG
4. Angular Momentum in Quantum Mechanics by A. R. EDMONDS
5. The Neutrino by JAMES S. ALLEN
7. Radiation Damage in Solids by D. BILLINGTON and J. CRAWFORD
8. Nuclear Structure by LEONARD EISENBUD and EUGENE P. WIGNER
9. The Physics of Elementary Particles by J. D. JACKSON
10. Invariance Principles and Elementary Particles by J. J. SAKURAI
11. Pion-Nucleon Scattering by ROBERT J. CENCE

PION-NUCLEON SCATTERING

BY

ROBERT J. CENCE

PRINCETON UNIVERSITY PRESS
PRINCETON, NEW JERSEY
1969

L. C. CARD: 66-11964
SBN: 691-08068-2

Printed in the United States of America
by Princeton University Press

PREFACE

This monograph is an up-to-date summary of the work of the last ten years in the field of pion-nucleon scattering. It is intended primarily for those outside this field who would like to acquaint themselves with the progress that has been made in the last decade. Any advanced graduate student with a thorough grounding in quantum mechanics should have little difficulty understanding this monograph. Also, workers just beginning to do research in pion-nucleon scattering will find this a useful introduction.

The text begins with two chapters of elementary scattering theory. Simple derivations are given for all the scattering formulas used. This includes the partial wave amplitudes for elastic scattering, expressions for inelastic and total cross sections, isotopic spin formulas, and the dispersion relations for the forward amplitude. There are special sections devoted to the optical theorem, the S-matrix, one-level resonance formulas, the Wigner condition, and the Pomeranchuk theorem. The use of eigen-phase shifts is described. A discussion of the A and R rotation parameters is given.

There is then a series of chapters which discuss total cross sections, the forward amplitude, elastic scattering and polarization, inelastic scattering, and photoproduction. A separate chapter discusses phase shift analysis. Experimental methods are described, results are displayed, and their interpretation given. Virtually all the experimental work below 2 GeV is either discussed or summarized. A complete set of references to all recent work is given at the end of each chapter.

Finally, the last chapter discusses the various models that have been proposed to describe the experimental results. These are the cut-off model, SU_3 theory, Regge poles, and the bootstrap idea. Unresolved problems that are worthy of future work are also indicated in this chapter.

The author would like to thank Professor Robert Hofstadter who encouraged this work and Professor Burton J. Moyer who helped outline it.

CONTENTS

PION-NUCLEON SCATTERING

CHAPTER 1

Introduction

1.1 Fundamental Role of the Pion-Nucleon Interaction

In nuclei the binding energy per particle is approximately constant. From this fact alone we can deduce that the forces between nucleons must have a finite range. This range can be roughly described by the formula:

$$r = r_0 A^{1/3}, r_0 \cong 1.5 \times 10^{-13} \text{ cm} \qquad (1.1)^1$$

It follows then that the range of the nucleon force is $\sim r_0$. Following Yukawa we assume that this force is due to the exchange of a particle of mass μ [Ref. 2]. The nucleon is pictured as a point particle of mass M constantly emitting particles of mass μ. Unless there is another nucleon nearby which can interact with the first by the absorption of one of these particles, they travel a finite distance R and return to the "core." The finite range R results from the fact that this process violates conservation of energy. Hence, it is allowed only within the confines of the uncertainty principle, $\Delta E \Delta t \geqslant \hbar$. The maximum range will be less than or equal to the range calculated when ΔE is minimum, namely μc^2 and when the velocity is a maximum, namely c.

$$\mu c^2 \cdot R/c \simeq \hbar \text{ where } R = \text{range} \qquad (1.2)$$

Thus $R \cong \hbar/\mu c$ = the Compton wavelength of the particle. If $R = r_0$, then we deduce a particle with rest energy $\sim$150 MeV.

Because of the great strength of nuclear forces, this particle must also interact strongly with nuclear particles. The $\pi^{\pm,\circ}$ mesons satisfy the above properties very nicely. As we shall see they interact strongly with nucleons, and their rest energies, agreeing with the above estimates, are [Ref. 3]:

$$\begin{aligned} \mu_{\pm} c^2 &= 139.577 \pm .013 \text{ MeV} \\ \mu_0 c^2 &= 134.973 \pm .014 \text{ MeV} \end{aligned} \qquad (1.3)$$

For this reason it is clear that this particle must play a fundamental role in the description of nuclear forces. The study of the interaction of the π meson with nucleons is thus one of the most fundamental problems in elementary particle physics.

The primary method of gaining information concerning this interaction is through scattering experiments. Because of their importance and because of the large amount of work that has been done, we will limit ourselves in this brief monograph to scattering phenomena involving pions and nucleons. Because of its close relationship with scattering, photoproduction will also be discussed.

[1] At high energies experiment indicates a smaller value $r_0 = 1.2 \times 10^{-13}$ cm [Ref. 1].

1.2 Importance of High Energies

We will be interested primarily in pion-nucleon (π-N) scattering at high energies. There are two reasons for this. First, in order to explore the details of the π-nucleon interaction, it is necessary that the wavelength of the two particles in the barycentric system be small. (Each particle will have the same momentum of course.) By the uncertainty principle the position of a particle is indeterminate by an amount of the order of its wavelength. Since the uncertainty in momentum in any one direction can at most be of the order of the total momentum, we have

$$\Delta x \sim \hbar/p_B = \lambda\!\!\!^{-}_B \tag{1.4}$$

where p_B is the momentum in the barycentric system. Thus, the particle can be used to probe details of the interaction which have a size no smaller than $\lambda\!\!\!^{-}_B$.

Another way to state the argument is as follows: the relative distance between the two particles will have an uncertainty determined by the uncertainty in the relative momentum of the two particles. This latter will be uncertain by a maximum given by the difference between the initial and final relative momenta, that is, the momentum transfer. Thus $\Delta x \sim \hbar/p_T$ where now p_T is the momentum transfer between the two particles. The minimum value of p_T is 0, the maximum value is $2p_B$, and thus the average is $\sim p_B$. The minimum distance, the impact parameter, must be $\geqslant \Delta x$. Thus again we can only probe details with size $\geqslant \lambda\!\!\!^{-}_B$.

To get an idea of the kinds of energies that concern us, we will calculate the wavelength in the barycentric system.

Let ω_L, k_L, E_L, p_L be the pion and nucleon total energy and momentum in the lab, respectively. The subscript B will be used to refer to these same quantities in the barycentric system. Fig. 1.1 illustrates the scattering in these two coordinate systems.

It is easily shown that $\omega p + kE$ is an invariant with respect to Lorentz transformations along the beam axis. Then we can write

$$0 + k_L M = \omega_B p_B + k_B E_B = p_B(\omega_B + E_B)$$

Hence,

$$k_B = \frac{k_L M}{\omega_B + E_B} \tag{1.5}$$

Since the square of the four momentum-energy vector is invariant, we have

$$(M + \omega_L)^2 - k_L{}^2 = (\omega_B + E_B)^2 \tag{1.6}$$

Hence, from (1.5) and (1.6),

$$k_B = \frac{k_L}{\sqrt{1 + \dfrac{2\omega_L}{M} + \left(\dfrac{\mu}{M}\right)^2}} \tag{1.7}$$

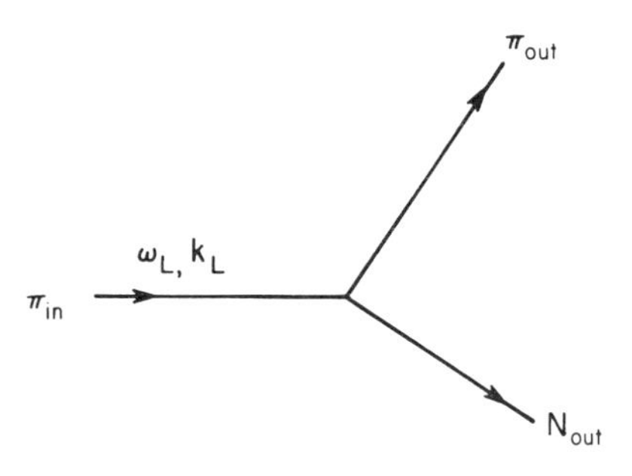

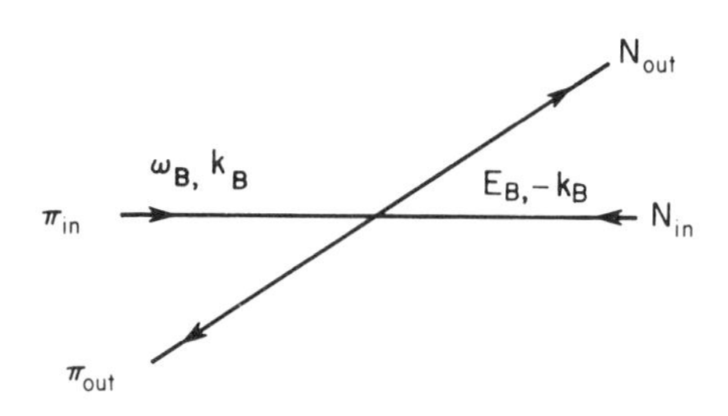

FIG. 1.1 A pion-nucleon elastic scatter as seen in the laboratory and barycentric coordinate systems.

and

$$\lambdabar_B = \lambdabar_L \sqrt{1 + 2\frac{\omega_L}{M} + \left(\frac{\mu}{M}\right)^2} \tag{1.8}$$

In Table 1.1, $\lambdabar_B$ is displayed at several energies for $\pi + N \to \pi + N$.

TABLE 1.1

KE (MeV)	$\lambdabar_B$ (Fermis)	$R/\lambdabar_B$	ℓ (From eq. 1.9)
150	1.0	1.41	1
370	0.58	2.45	2
660	0.41	3.47	3
1,010	0.32	4.47	4
1,500	0.26	5.48	5
2,000	0.22	6.48	6
2,500	0.19	7.49	7
5,000	0.135	10.50	10

KE = kinetic energy of incident pion in the lab.
$\lambdabar_B$ = wavelength in the barycentric system.
$R = \frac{\hbar}{\mu c} = 1.41 \times 10^{-13}$ cm.

It is evident from Table 1.1 that if the π-N interaction is to be explored at distances which are a small fraction of a pion Compton wavelength, experiments must be performed at energies of several hundreds of MeV and beyond.

However, there is also another reason for exploring the interaction at high energies. The finite range of the force puts an upper limit on the lever arm which defines the angular momentum. Thus, if we wish to explore the interaction at higher angular momenta, we must have high energies. The maximum angular momentum will occur when the lever arm is equal to R, the range of the interaction. That is,

$$Rp_{\mathrm{B}} \geqslant \sqrt{\ell(\ell+1)}\,\hbar$$

and hence

$$R/\lambda\!\!\!^{-}_{\mathrm{B}} \geqslant \sqrt{\ell(\ell+1)} \tag{1.9}$$

This relation is not very meaningful unless $\ell \geqslant 2$ since the uncertainty relation (1.4) must also be satisfied. The quantity $R/\lambda\!\!\!^{-}_{\mathrm{B}}$ is displayed in Table 1.1 at several energies. It is clear that to explore angular momenta > 1 we need energies greater than 200 MeV.

1.3 Isotopic Spin

Experimentally we know that the π meson exists in three charged states. In units of the electron charge, it has the values +1, 0, − 1. Formally, this can be described by an operator which has the above eigenvalues. These are the same eigenvalues as the z component of the angular momentum for total angular momentum 1 (in units of $\hbar$). Thus, to describe different charged states an operator which has the same formal properties as the angular momentum operator can be used. Thus, the formalism developed in quantum mechanics for angular momentum can be carried over unchanged for the description of isotopic spin. For the pion the isotopic spin eigenfunctions will be formally identical to the angular momentum eigenfunctions for angular momentum 1.

The eigenvalues of the charge operator are called isotopic spin. For the isotopic spin operators we write τ_1, τ_2, τ_3. Assume τ_3 and τ^2 are diagonal with eigenvalues T_3, T respectively. Then for the π meson

$$Q = T_3 \quad \text{(pions)} \tag{1.10}$$

where Q is the charge in units of e. The nucleon exists in only two charged states and hence will have isotopic spin $\frac{1}{2}$, a doublet. Then

$$Q = T_3 + \tfrac{1}{2} \quad \text{(nucleons)} \tag{1.11}$$

To describe the charge states of a pion and a nucleon a total isotopic spin can be defined which is the vector sum of the isotopic spins of the two particles. The rules for addition are the same as those for addition of angular momenta. This means that there will be two possible total isotopic spins for the

π-N system, $T = \frac{1}{2}$, and $T = \frac{3}{2}$. In general a given π-N state, π^+-neutron for example, will not be in a single isotopic spin state but will instead be a linear combination of $T = \frac{1}{2}$ and $T = \frac{3}{2}$. The coefficients will be just the Clebsch-Gordan coefficients for the addition of angular momentum.

For completeness we write down the π-N state functions $|\pi N\rangle$ in terms of the isotopic spin state functions $|T, T_3\rangle$.

$$\begin{aligned}
|\pi^+p\rangle &= |\tfrac{3}{2}, \tfrac{3}{2}\rangle \\
|\pi^+n\rangle &= \sqrt{\tfrac{1}{3}}\,|\tfrac{3}{2}, \tfrac{1}{2}\rangle + \sqrt{\tfrac{2}{3}}\,|\tfrac{1}{2}, \tfrac{1}{2}\rangle \\
|\pi^\circ p\rangle &= \sqrt{\tfrac{2}{3}}\,|\tfrac{3}{2}, \tfrac{1}{2}\rangle - \sqrt{\tfrac{1}{3}}\,|\tfrac{1}{2}, \tfrac{1}{2}\rangle \\
|\pi^\circ n\rangle &= \sqrt{\tfrac{2}{3}}\,|\tfrac{3}{2}, -\tfrac{1}{2}\rangle + \sqrt{\tfrac{1}{3}}\,|\tfrac{1}{2}, -\tfrac{1}{2}\rangle \\
|\pi^-p\rangle &= \sqrt{\tfrac{1}{3}}\,|\tfrac{3}{2}, -\tfrac{1}{2}\rangle - \sqrt{\tfrac{2}{3}}\,|\tfrac{1}{2}, -\tfrac{1}{2}\rangle \\
|\pi^-n\rangle &= |\tfrac{3}{2}, -\tfrac{3}{2}\rangle .
\end{aligned} \tag{1.12}$$

Thus far we have adopted merely an alternative formalism for describing charge states. There are six $|\pi N\rangle$ states and six $|T, T_3\rangle$ states. No physical restrictions have been introduced.

There is by now overwhelming evidence for the hypothesis that the strong interactions are charge independent. In particular this includes nuclear forces and the π-N interaction. For a precise experiment which tests this hypothesis see Ref. 4. In Chapter 3 some of the evidence for charge independence will be examined.

In addition to the purely nuclear interaction there is also a Coulomb interaction between charged pions and charged nucleons. In general this is so small compared to the nuclear interaction that it is at most a small perturbation. Often in fact it can be ignored. Where necessary we will show how it can be taken into account.

The hypothesis of charge independence means that the π-N interaction is independent of T_3. That is, it is independent of the orientation of the isotopic spin vector in isotopic spin space. Since isotopic spin space is isotropic with respect to strong interactions, then like angular momentum in a central field, *the total isotopic spin is conserved.*

Because in quantum mechanics as well as in classical physics the equations of motion can be solved for at most two bodies, elastic scattering will play a central role in this subject. There are ten possible reactions including the phenomenon of charge exchange:

$$\pi^+ p \longrightarrow \pi^+ p \qquad \text{(i)}$$

$$\pi^- p \longrightarrow \pi^- p \qquad \text{(ii)}$$

$$\pi^- p \longrightarrow \pi^\circ n \qquad \text{(iii)}$$

$$\pi^+ n \longrightarrow \pi^+ n \qquad \text{(iv)}$$

$$\begin{array}{ll} \pi^+ n \longrightarrow \pi^\circ p & \text{(v)} \\ \pi^- n \longrightarrow \pi^- n & \text{(vi)} \\ \pi^\circ p \longrightarrow \pi^\circ p & \text{(vii)} \\ \pi^\circ n \longrightarrow \pi^\circ n & \text{(viii)} \\ \pi^\circ n \longrightarrow \pi^- p & \text{(ix)} \\ \pi^\circ p \longrightarrow \pi^+ n & \text{(x)} \end{array} \tag{1.13}$$

Reactions (ix) and (x) are related to (iii) and (v) by the principle of time reversal invariance. Reactions (vii) and (viii) are not feasible experimentally due to the short lifetime of the π° meson $(= 1.8 \times 10^{-16}$ sec). Reactions (iv), (v) and (vi) are not very useful because they involve using neutron targets. Thus, only (i), (ii), and (iii) are readily amenable to experiment.

The cross sections for all of the above reactions will be equal to the square of some matrix element,

$$\sigma = |\langle \pi N | M | \pi N \rangle|^2 \tag{1.14}$$

By the hypothesis of charge independence T is conserved and the matrix element is independent of T_3. Thus, all ten of the above reactions depend on only two amplitudes.

$$\left\langle \tfrac{1}{2} \middle| M \middle| \tfrac{1}{2} \right\rangle = F_1$$

and

$$\left\langle \tfrac{3}{2} \middle| M \middle| \tfrac{3}{2} \right\rangle = F_3 \tag{1.15}$$

where T_3 has been suppressed and M is a scattering operator.

From (1.12), (1.14), and (1.15) the cross sections for reactions (i), (ii), and (iii) can be expressed as,

$$\begin{aligned} \sigma(\pi^+ p \longrightarrow \pi^+ p) &= |F_3|^2 \\ \sigma(\pi^- p \longrightarrow \pi^- p) &= \left|\tfrac{1}{3} F_3 + \tfrac{2}{3} F_1\right|^2 \\ \sigma(\pi^- p - \pi^\circ n) &= \tfrac{2}{9}\left|F_3 - F_1\right|^2 \end{aligned} \tag{1.16}$$

Simple relations can also be derived between the $\pi^\pm p$ total cross section, σ_{tot}^{+-}, and the total cross sections for scattering in the $T = \frac{1}{2}$ and $T = \frac{3}{2}$ isotopic spin states, $\sigma_{\text{tot}}^{(1)}$ and $\sigma_{\text{tot}}^{(3)}$ respectively. Thus

$$\sigma_{\text{tot}}^- = |\sum_f \langle f | M | \pi^- p \rangle|^2 \tag{1.17}$$

where the sum runs over all possible final states f. We can write (1.17) as

$$\sigma_{\text{tot}}^- = \sum_f \sum_{f'} \langle \pi^- p | M^+ | f \rangle \langle f' | M | \pi^- p \rangle \tag{1.18}$$

or

$$\sigma_{\text{tot}}^- = \langle \pi^- p | N | \pi^- p \rangle$$

where

$$N = \sum_f \sum_{f'} M^+|f\rangle\langle f'|M \tag{1.19}$$

is an operator which conserves isotopic spin. Thus from (1.12) and (1.18),

$$\sigma_{\text{tot}}^- = \tfrac{1}{3}\langle \tfrac{3}{2}|N|\tfrac{3}{2}\rangle + \tfrac{2}{3}\langle \tfrac{1}{2}|N|\tfrac{1}{2}\rangle \tag{1.20}$$

or

$$\sigma_{\text{tot}}^- = \tfrac{1}{3}\sigma_{\text{tot}}^{(3)} + \tfrac{2}{3}\sigma_{\text{tot}}^{(1)} \tag{1.21}$$

and of course

$$\sigma_{\text{tot}}^+ = \langle \tfrac{3}{2}|N|\tfrac{3}{2}\rangle \tag{1.22}$$

so

$$\sigma_{\text{tot}}^+ = \sigma_{\text{tot}}^{(3)} \tag{1.23}$$

We can solve for $\sigma_{\text{tot}}^{(1)}$ obtaining

$$\sigma_{\text{tot}}^{(1)} = \tfrac{3}{2}\sigma_{\text{tot}}^- - \tfrac{1}{2}\sigma_{\text{tot}}^+ \tag{1.24}$$

Below, the first inelastic channel (1.23) and (1.24) become

$$\begin{aligned} \sigma^{(3)} &= \sigma^+ \\ \sigma^{(1)} &= \tfrac{3}{2}(\sigma^- + \sigma^\circ) - \tfrac{1}{2}\sigma^+ \end{aligned} \tag{1.25}$$

where, by σ^+, σ^- and σ° are meant the cross sections for reactions (i), (ii), and (iii) in (1.13). We have recognized that σ_{tot}^- includes charge exchange. The relation (1.25) holds even if the inelastic scattering is not 0. This is easily seen by substituting expressions (1.16) directly into the right-hand side of the second equation of (1.25).

1.4 Spin and Parity of the π-Meson

We mention briefly the important properties of spin and parity of the π meson. Within a few years after their discovery in 1947, it was demonstrated conclusively that the π mesons have zero spin and odd parity. Their state functions are thus pseudo-scalars. For an extensive discussion of the experimental evidence regarding these properties, see Ref. 5.

1.5 Conservation Laws

The π-N interaction is among the strongest interactions known. One view is that its strength is due to the fact that it conserves all the quantities which are known to have invariance properties. Energy and momentum are of course conserved. Also, charge and baryon number are conserved by all the known interactions. There are, however, some quantities which are conserved only in the strong interactions such as the π-N interaction. These include isotopic spin as previously discussed, strangeness, and parity. The strong interaction is also invariant with respect to charge conjugation and time reversal. Isotopic spin rotational invariance allows us, for example in (1.13), to equate reactions

(i) and (vi) and reactions (ii) and (iv). Invariance under time reversal allows us in (1.13) to equate reactions (iii) and (v) to (ix) and (x), respectively.

REFERENCES

1. International Congress on Nuclear Sizes and Density Distributions held at Stanford University 1957, *Rev. Mod. Phys. 30,* 412 (1958).
2. H. Yukawa, *Proceedings Physics-Math Society, Japan 17,* 48 (1935).
3. R. E. Shafer, *Phys. Rev. 163,* 1451 (1967); I. M. Vasilevsky, V. V. Vishnyakov, A. F. Dunaitsev, Yu. D. Prokoshkin, V. I. Rykalin, and A. A. Tyapkin, *Phys. Letters 23,* 281 (1966); J. B. Czirr, *Phys. Rev. 130,* 341 (1963).
4. J. A. Poirier and M. Pripstein, *Phys. Rev. 130,* 1171 (1963).
5. H. A. Bethe and F. De Hoffmann, Chapter 28, *Mesons and Fields,* Vol. II, Row, Peterson and Co. (1955).

CHAPTER 2

Elementary Scattering Theory

In this chapter, some of the formulas for $\pi - N$ scattering will be derived. These formulas will be functions of parameters called phase shifts. They can be interpreted as the shift in phase of the wave function at a large distance from the scattering center and resulting from the scattering process. There will be a separate phase shift for scattering in each state of total angular momentum and parity. Thus, the phase shifts have a simple and basic physical meaning.

The derivations in this chapter will be non-relativistic. Nevertheless, the results are all true relativistically also. For a discussion of this point, see Ref. 1.

2.1 Elastic Scattering: Spin 0

We first summarize the results for elastic scattering of two spin 0 particles [Ref. 2]. It will be assumed, unless stated otherwise, that we are working in the barycentric system where the total momentum is 0:

$$\Sigma \vec{p}_i = 0 \tag{2.1}$$

In this system, the Schrödinger equation reduces to the soluble one-body problem with a reduced mass m_r given by

$$\frac{1}{m_r} = \frac{1}{m_1} + \frac{1}{m_2} \tag{2.2}$$

where m_1 and m_2 are the masses of the scattered and scattering particles. The asympototic wave function which describes scattering in the barycentric system is given by

$$\psi(\vec{r}) = e^{ikz} + F(\theta, k)\frac{e^{ikr}}{r} \tag{2.3}$$

where k is the wave number. The first term in (2.3) represents the incoming wave propagating along the positive z-axis. The second term represents the scattered wave as an outgoing spherical wave. The quantity $F(\theta, k)$ is called the scattering amplitude. The differential cross section for elastic scattering is given by

$$\frac{d\sigma(\theta, k)}{d\Omega} = |F(\theta, k)|^2 \tag{2.4}$$

The general expression for $F(\theta, k)$ in terms of partial waves is

$$F(\theta, k) = \frac{1}{k}\sum_{\ell=0}^{\infty}(2\ell + 1)a_\ell(k)P_\ell(\cos\theta) \tag{2.5}$$

where we define

$$a_\ell(k) = \frac{\exp\{2i\delta_\ell(k)\} - 1}{2i} \tag{2.6}$$

The $a_\ell(k)$ defined in (2.6) is called the partial wave amplitude. Each term in the sum represents scattering in the state with angular momentum ℓ. The $\delta_\ell(k)$ are the phase shifts. Asymptotically the wave function in the state with angular momentum ℓ is given by:

$$\psi_\ell(r) \propto \sin\{kr - \tfrac{1}{2}\ell\pi + \delta_\ell(k)\} \quad \text{as } r \longrightarrow \infty \tag{2.7}$$

Equation (2.7) assumes that the force responsible for the scattering has a finite range, or more precisely, falls off faster than $1/r^2$. This, as was shown in Chapter 1, is certainly true for the nuclear force.

2.2 Elastic Scattering: Spin $\frac{1}{2}$

In the scattering of a spin $\frac{1}{2}$ particle by a spin 0 particle, there is the possibility that the scattering process may produce polarization. The wave function must include the spin $\frac{1}{2}$ functions

$$\psi(\vec{r}) = \left(e^{ikz} + F(\theta,\phi,k)\frac{e^{ikr}}{r}\right)\chi \tag{2.8}$$

where χ is the spin function of the incident wave. Now $F(\theta, \phi, k)$ is a 2×2 matrix. Due to the spin of the incoming particle it may depend on the azimuthal angle ϕ.

We assume that the z-axis is the axis of quantization for angular momentum. The spin function will be a linear combination of the two spin eigenfunctions:

$$\alpha = \begin{pmatrix} 1 \\ 0 \end{pmatrix}, \quad \beta = \begin{pmatrix} 0 \\ 1 \end{pmatrix} \tag{2.9}$$

These are eigenfunctions of σ^2 and σ_z where

$$\sigma_x = \begin{pmatrix} 0 & 1 \\ 1 & 0 \end{pmatrix}, \quad \sigma_y = \begin{pmatrix} 0 & -i \\ i & 0 \end{pmatrix}, \quad \sigma_z = \begin{pmatrix} 1 & 0 \\ 0 & -1 \end{pmatrix} \tag{2.10}$$

The α is the spin wave function for spin directed along the $+z$-direction and β that for spin along the $-z$-direction. The wave function (2.8) is in a single spin state and hence describes a fully polarized incident wave where polarization is defined,

$$\vec{P} = \frac{\langle\psi|\vec{\sigma}|\psi\rangle}{\langle\psi|\psi\rangle} \tag{2.11}$$

For a given orbital angular momentum ℓ, the total angular momentum j

can have two values, $\ell + \frac{1}{2}$ and $\ell - \frac{1}{2}$. The scattering amplitude must now be generalized to allow for different scattering in these two states. To indicate this, we write $a_{\ell j}(k)$ for the partial wave amplitude defined in (2.6) and replace a_ℓ in (2.5) by

$$a_{\ell,\ell+1/2}(k)\,\Lambda_{\ell+} + a_{\ell,\ell-1/2}\,(k)\,\Lambda_{\ell-} \tag{2.12}$$

where

$$\begin{aligned}
&\Lambda_{\ell+}\psi_j(j = \ell + \tfrac{1}{2}) = \psi_j(j = \ell + \tfrac{1}{2})\\
&\Lambda_{\ell+}\ \psi_j(j = \ell - \tfrac{1}{2}) = 0\\
&\Lambda_{\ell-}\psi_j(j = \ell - \tfrac{1}{2}) = \psi_j(j = \ell - \tfrac{1}{2})\\
&\Lambda_{\ell-}\psi_j(j = \ell + \tfrac{1}{2}) = 0
\end{aligned} \tag{2.13}$$

and

$$\Lambda_{\ell+} + \Lambda_{\ell-} = 1. \tag{2.14}$$

If $a_{\ell,\ell+1/2}(k) = a_{\ell,\ell-1/2}(k) = a_\ell(k)$, then by (2.14)

$$a_{\ell,\ell+1/2}(k)\Lambda_{+\ell} + a_{\ell,\ell-1/2}(k)\Lambda_{-} = a_\ell(k)\,(\Lambda_{+} + \Lambda_{-}) = a_\ell(k)$$

This takes us back to the spin zero case. The Λ_+, Λ_- are operators which project out the states $j = \ell + \frac{1}{2}$ and $j = \ell + \frac{1}{2}$, respectively.

It is easily shown that[1]

$$\begin{aligned}
\Lambda_{\ell+} &= \frac{\ell + 1 + \vec{\ell}\cdot\vec{\sigma}}{2\ell + 1}\\
\Lambda_{\ell-} &= \frac{\ell - \vec{\ell}\cdot\vec{\sigma}}{2\ell + 1}
\end{aligned} \tag{2.15}$$

If one uses (2.15) and (2.12), the scattering amplitude can be written

$$\begin{aligned}
F(\theta,\phi,k) &= \frac{1}{k}\sum_{\ell=0}^{\infty}\ [(\ell + 1)a_{\ell,\ell+1/2}(k) + \ell a_{\ell,\ell-1/2}(k)]\ P_\ell(\cos\theta)\\
&+ \frac{1}{k}\sum_{\ell=0}^{\infty}\ [a_{\ell,\ell+1/2}(k) - a_{\ell,\ell-1/2}(k)]\ \{\vec{\ell}P_\ell(\cos\theta)\}\cdot\vec{\sigma}
\end{aligned} \tag{2.16}$$

since $\vec{\sigma}$ does not operate on $P_\ell(\cos\theta)$. It operates on the spin function χ in (2.8).

Now,

$$\vec{\ell} = \frac{1}{i}\vec{r}\times\vec{\nabla} = \frac{1}{i}\vec{r}\times\vec{n}_\theta\,\frac{d}{rd\theta} = \frac{1}{i}\,\vec{n}_r\times\vec{n}_\theta\,\frac{d}{d\theta} \tag{2.17}$$

[1]To show this write $\vec{\ell}\cdot\vec{\sigma} = (\vec{\ell} + \frac{1}{2}\vec{\sigma})^2 - \ell^2 - \frac{1}{4}\sigma^2 = j^2 - \ell^2 - \frac{1}{4}\sigma^2$.

since $P_\ell(\cos\theta)$ does not depend on ϕ. Here $\vec{n}_r$ and $\vec{n}_\theta$ are unit vectors in the r and θ directions, respectively.

Thus,

$$\begin{aligned}\vec{\ell} P_\ell(\cos\theta) &= \frac{1}{i}\,\vec{n}_r \times \vec{n}_\theta \sin\theta \frac{d}{d\cos\theta} P \;(\cos\theta) \\ &= -\frac{1}{i}\,\vec{n}_\perp P_\ell^1(\cos\theta)\end{aligned} \tag{2.18}$$

where $\vec{n}_\perp = \vec{n}_\theta \times \vec{n}_r$

We can now write (2.16) as

$$F(\theta,\phi,k) = f(\theta,k) - g(\theta,k)\vec{n}_\perp \cdot \vec{\sigma} \tag{2.19}$$

where

$$\left.\begin{aligned} f(\theta,k) &= \frac{1}{k}\sum_{\ell=0}^{\infty}\{(\ell+1)\,a_{\ell,\ell+1/2}(k) + a_{\ell,\ell-1/2}(k)\}\,P(\cos\theta) \\ &= \sum_{\ell=0}^{\infty}\sum_{j=|\ell-1/2|}^{\ell+1/2} (j+\tfrac{1}{2})\,a_{\ell j}(k)\,P_\ell(\cos\theta) \\ \text{and}\quad g(\theta,k) &= \frac{1}{ik}\sum_{\ell=0}^{\infty}\{a_{\ell,\ell+1/2}(k) - a_{\ell,\ell-1/2}(k)\}\,P_\ell^1(\cos\theta) \end{aligned}\right\} \tag{2.20}$$

Since $f(\theta)\,\chi$ has the *same* spin as the initial state, $f(\theta)$ is called the non-spin flip amplitude. On the other hand, $\vec{n}\cdot\vec{\sigma}$ is an anti-diagonal matrix. Thus, operating on χ, we have $\alpha \to \beta, \beta \to \alpha$, and hence

$$g(\theta,k)\vec{n}_\perp \cdot \vec{\sigma}$$

has a spin *opposite* to the initial state. Hence, it is called the spin-flip amplitude.

We now want to calculate the differential cross section and polarization of the scattered particles. We will assume at first that the incident particles are *fully polarized* in the direction θ_i, ϕ_i. From the definition (2.11) it is easy to show that

$$\chi = \begin{pmatrix} \cos\dfrac{\theta_i}{2}\exp\left\{-\dfrac{i\phi_i}{2}\right\} \\ \sin\dfrac{\theta_i}{2}\exp\left\{\dfrac{i\phi_i}{2}\right\} \end{pmatrix} \tag{2.21}$$

represents a wave fully polarized in the direction θ_i, ϕ_i.

From (2.4),

$$\frac{d\sigma}{d\Omega} = [F(\theta,\phi,k)\,\chi\,]^{\dagger}\,[F(\theta,\phi,k)\,\chi\,] \tag{2.22}$$

The use of the Hermitian conjugate, indicated by the †, in (2.22) is an obvious generalization when $F(\theta,\phi,k)$ is a matrix.

Now $\frac{d\sigma}{d\Omega}$ depends on $\theta,\phi,\theta_i,\phi_i$, and k.

Using (2.19) and (2.21) we have

$$F(\theta,\phi,\theta_i,\phi_i,k)\,\chi = \begin{pmatrix} f\cos\frac{\theta_i}{2}\exp\left\{-\frac{i\phi_i}{2}\right\} + i\,g\sin\frac{\theta_i}{2}\exp\left\{i\left(\frac{\phi_i}{2}-\phi\right)\right\} \\ f\sin\frac{\theta_i}{2}\exp\left\{\frac{i\phi_i}{2}\right\} - i\,g\cos\frac{\theta_i}{2}\exp\left\{-i\left(\frac{\phi_i}{2}-\phi\right)\right\} \end{pmatrix} \tag{2.23}$$

Then (2.22) gives,

$$\frac{d\sigma}{d\Omega} = |f(\theta,k)|^2 + |g(\theta,k)|^2 + 2(\vec{n}_\perp\cdot\vec{P}_i)\ \mathrm{Re}[f^*(\theta,k)\,g(\theta,k)] \tag{2.24}$$

For the polarization of the scattered particles we write from definition (2.11),

$$\vec{P} = \frac{\langle\psi_s\,|\,\vec{\sigma}\,|\,\psi_s\rangle}{\langle\psi_s\,|\,\psi_s\rangle} \tag{2.25}$$

where,

$$\psi_s = F(\theta,\phi,\theta_i,\phi_i,k)\,\chi\frac{e^{ikr}}{r} \tag{2.26}$$

Substituting (2.23) and (2.26) into (2.25) we find, after a little tedious but elementary algebra,

$$\begin{aligned}\frac{d\sigma}{d\Omega}\cdot\vec{P}(\theta,\phi,\theta_i,\phi_i,k) = {} & 2\,\mathrm{Re}\,[f^*(\theta,k)\,g\,(\theta,k)]\,\vec{n}_\perp - \\ & 2\,\mathrm{Im}\,[f^*(\theta,k)g\,(\theta,k)]\,\vec{n}_\perp\times\vec{P}_i + (|f(\theta,k)|^2 - |g(\theta,k)|^2)\vec{P}_i + \\ & 2|g\,(\theta,k)|^2\,(\vec{n}_\perp\cdot\vec{P}_i)\,\vec{n}_\perp\end{aligned} \tag{2.27}$$

where $\frac{d\sigma}{d\Omega}$ is given by expression (2.24) and $\vec{n}_\perp$ is a unit vector perpendicular to the scattering plane. It is defined,

$$n_\perp = \frac{\vec{k}_{\mathbf{out}}\times\vec{k}_{\mathrm{in}}}{|\vec{k}_{\mathbf{out}}\times\vec{k}_{\mathrm{in}}|} \tag{2.28}$$

where $\vec{k}_{\text{in}}$ and $\vec{k}_{\text{out}}$ are the momentum vectors of the incoming and scattered particles respectively, and $\vec{P}_i$ is the polarization of the incoming particles. This is the same $\vec{n}_\perp$ that appears in eq. 2.18.

Let us now calculate the differential cross section and polarization of the scattered particles when the incoming particles are completely *unpolarized*. To determine first the differential cross section we use expression (2.22) averaging *incoherently* over the two spin eigen-states α and β defined in (2.9). Thus,

$$\frac{d\sigma}{d\Omega} = \frac{1}{2}\sum_{j=1}^{2} [F(\theta,\phi,k)\chi_i]^\dagger \, [F(\theta,\phi,k)\chi_i] \tag{2.29}$$

where we set $\chi_i = \alpha$ and $\chi_2 = \beta$. We find

$$\frac{d\sigma}{d\Omega} = |f(\theta,k)|^2 + |g(\theta,k)|^2 \equiv I_0 \tag{2.30}$$

In a similar way we write

$$\vec{P} = \frac{1}{2}\sum_{\text{spins}} \frac{\langle\psi_s|\vec{\sigma}|\psi_s\rangle}{\langle\psi_s|\psi_s\rangle} \tag{2.31}$$

where ψ_s is given by (2.26). We find

$$\vec{P} = 2\,\text{Re}\,[f^*(\theta,k)\,g(\theta,k)]\,\vec{n}_\perp \tag{2.32}$$

We note that the only difference between expressions (2.30), (2.32) and (2.24), (2.27) is that terms linear in $\vec{P}_i$ have disappeared. Thus, it is reasonable to assume that (2.24) and (2.27) hold also for *partially polarized incoming beams of particles* where now $|\vec{P}_i| < 1$ [Ref. 3]. We will apply these formulas to several special cases in Chapter 5.

Using expressions (2.16) and (2.20) one can easily integrate (2.24) over solid angle to obtain the total elastic cross sections. The cross terms and the term in $\vec{P}_i$ cancel out and one obtains:

$$\begin{aligned}\sigma_{\text{el}}(k) &= \pi\lambda\!\!\!^{-2}\sum_{\ell=0}^{\infty}\{(\ell+1)|1-\exp\{2i\delta_{\ell,\ell+1/2}(k)\}|^2 \\ &\qquad + \ell|1-\exp\{2i\delta_{\ell,\ell-1/2}(k)\}|^2\} \\ &= \pi\lambda\!\!\!^{-2}\sum_{\ell=0}^{\infty}\sum_{j=|\ell-1/2|}^{\ell+1/2}(j+\tfrac{1}{2})|1-\exp\{2i\delta_{\ell j}(k)\}|^2\end{aligned} \tag{2.33}$$

where $\lambda\!\!\!^{-} = 1/k$.

2.3 Coulomb Scattering

It is convenient to account explicitly for Coulomb scattering in the expression for $F(\theta, \phi, k)$ defined in (2.19). In this section we will summarize without derivation the changes needed to do this. Write $f_{\text{coul}}(\theta, k)$, $g_{\text{coul}}(\theta, k)$ for the terms which must be added to the non-spin flip and spin-flip amplitudes defined by eq. (2.20). Then

$$f_{\text{coul}}(\theta, k) = -\frac{\bar{\lambda}\alpha}{2\sin^2(\theta/2)} \exp\{-i\alpha \ln[\sin^2(\theta/2)]\} \quad \text{for } \pi^{\pm}p \longrightarrow \pi^{\pm}p,$$

$$f_{\text{coul}}(\theta,k) = 0 \quad \text{for } \pi^- p \longrightarrow \pi^0 n \tag{2.34}$$

and

$$g_{\text{coul}}(\theta, k) = 0 \quad \text{all charge states}$$

where $\alpha = \pm \dfrac{e^2}{\hbar v}$, v = relative velocity of incident and target particles, and plus sign for $\pi^+ p$ and minus sign for $\pi^- p$. In addition the partial wave amplitudes $a_{\ell j}(k)$ must be multiplied by a phase factor,

$$\begin{aligned} a_{\ell j}(k) &\longrightarrow \exp\{2i\phi_\ell\}\, a_{\ell j}(k) \quad \text{for } \pi^{\pm}p \longrightarrow \pi^{\pm}p \\ a_{\ell j}(k) &\longrightarrow \exp\{i\phi_\ell\}\, a_{\ell j}(k) \quad \text{for } \pi^- p \quad \pi^0 p \end{aligned} \tag{2.35}$$

where

$$\phi_0 = 0$$

and

$$\phi_\ell = \sum_{n=1}^{\ell} \tan^{-1}\left(\frac{\alpha}{n}\right), \quad \ell \geqslant 1 \tag{2.36}$$

The ϕ_ℓ are called the Coulomb phase shifts. For charge exchange scattering the ϕ_ℓ are calculated using the minus sign in the definition of α. The above formulas are all non-relativistic. For a derivation see pp. 114–121 of Ref. 2. The modifications necessary for charge exchange scattering are due to Sorensen [Ref. 4]. It is reasonable that the Coulomb phase shifts in charge exchange scattering should be half as large. This is because the Coulomb field is active during only half the scattering.

Relativistic corrections to Coulomb scattering have been calculated [Ref. 5, 6]. Expressions (2.35) and (2.36) remain the same. However, the expressions for $f_{\text{coul}}(\theta, k)$ and $g_{\text{coul}}(\theta, k)$ given in (2.34) are no longer correct. In particular $g_{\text{coul}}(\theta, k) \neq 0$.

We note in passing that the Coulomb phase shifts defined in (2.36) are small except at very low energies. For pion kinetic energies greater than 20 MeV we have $\phi_\ell < 1$ deg. for all ℓ.

2.4 Optical Theorem

If we take the wave function (2.8) and calculate the total outward flux,

$$S = \frac{\hbar}{2im} \int [\psi^* \vec{\nabla} \psi - \psi \vec{\nabla} \psi^*] \, r^2 \sin\theta \, d\theta \, d\phi \qquad (2.37)$$

we find

$$S = v \left[\sigma_{el} - \frac{4\pi}{k} \operatorname{Im} F(0, k) \right] \qquad (2.38)$$

where v is the relative velocity and $F(0, k)$ is the forward scattering amplitude.

If there were no inelastic scattering, we would have $S = 0$, since then the incoming flux would equal the outgoing flux. If there is inelastic scattering, there is a net loss of particles with momentum k, and hence

$$S = -v\sigma_{in} = v \left[\sigma_{el} - \frac{4\pi}{k} \operatorname{Im} F(0,k) \right] \qquad (2.39)$$

where σ_{in} is the inelastic cross section. Since the total cross section $\sigma_{tot}(k) = \sigma_{el}(k) + \sigma_{in}(k)$ we have finally,

$$\sigma_{tot}(k) = \frac{4\pi}{k} \operatorname{Im} [F(0, k)] \qquad (2.40)$$

This last result is usually referred to as the optical theorem.

2.5 Total Cross Section

If we use expressions (2.19) and (2.20) for the scattering amplitude in (2.40), we can express the total cross section in terms of partial waves. We first note that the spin-flip amplitude $g(\theta)$ vanishes at 0 deg. One then finds that,

$$\sigma_{tot}(k) = 4\pi\lambda\!\!\!^{-2} \sum_{\ell=0}^{\infty} \{(\ell+1) \operatorname{Im} a_{\ell,\ell+1/2} + \ell \operatorname{Im} a_{\ell,\ell-1/2}\}$$

$$= 2\pi\lambda\!\!\!^{-2} \sum_{\ell=0}^{\infty} \sum_{j=|\ell-1/2|}^{\ell+1/2} (j + \tfrac{1}{2})(1 - \operatorname{Re} \exp\{2i\delta_{\ell j}(k)\}) \qquad (2.41)$$

2.6 Inelastic Cross Section

To calculate the inelastic cross section, we take $\sigma_{in}(k) = \sigma_{tot}(k) - \sigma_{el}(k)$.

Using (2.33) and (2.41) we obtain

$$\sigma_{in}(k) = \pi\lambda^2 \sum\{(\ell + 1)\,(1 - |\exp\{2i\delta_{\ell,\ell+1/2}(k)\}|^2) + \ell(1 - |\exp\{2i\delta_{\ell,\ell-1/2}(k)\}|^2)\} \tag{2.42}$$

From this last expression we see that the inelastic cross section will be different from zero only when one or more of the $\delta_{\ell j}(k)$ are *complex*. Further, since the $\delta_{\ell j}(k)$ are independent, we must in general have

$$|\exp\{2i\delta_{\ell j}(k)\}| \leqslant 1 \quad \text{for all } \ell, j \tag{2.43}$$

in order that the inelastic cross section be positive.

It is convenient to separate the real and imaginary parts of the phase shifts as follows: we redefine the partial wave amplitude

$$a_{\ell j}(k) = \frac{\eta_{\ell j}(k)\exp\{2i\delta_{\ell j}(k)\} - 1}{2i} \tag{2.44}$$

where now the $\delta_{\ell j}(k)$ are all real. The $\eta_{\ell j}(k)$ are called absorption parameters. From (2.43) we have

$$0 \leqslant \eta_{\ell j}(k) \leqslant 1 \tag{2.45}$$

We can now rewrite (2.33), (2.41), and (2.42) as follows:

$$\sigma_{el}(k) = \pi\lambda^2 \sum_{\ell=0}^{\infty} \sum_{j=|\ell-1/2|}^{\ell+1/2} (j + \tfrac{1}{2})\,[\{1 - \eta_{\ell j}(k)\}^2 + 4\eta_{\ell j}(k)\sin^2\delta_{\ell j}(k)] \tag{2.46}$$

$$\sigma_{tot}(k) = 2\pi\lambda^2 \sum_{\ell=0}^{\infty} \sum_{j=|\ell-1/2|}^{\ell+1/2} (j + \tfrac{1}{2})\,[1 - \eta_{\ell j}(k) + 2\eta_{\ell j}(k)\sin^2\delta_{\ell j}(k)] \tag{2.47}$$

$$\sigma_{in}(k) = \pi\lambda^2 \sum_{\ell=0}^{\infty} \sum_{j=|\ell-1/2}^{\ell+1/2} (j + \tfrac{1}{2})\,[1 - \{\eta_{\ell j}(k)\}^2] \tag{2.48}$$

Note that these expressions are all linear in the sum over ℓ and j and each term is positive. This is in contrast to expressions (2.24) and (2.27) for $\frac{d\sigma(\theta, k)}{d\Omega}$ and $P(\theta, k)$. This means that while interference between different partial waves is possible for the latter, none is possible for the former. The addition of more partial waves will always increase $\sigma_{el}(k)$, $\sigma_{tot}(k)$, and $\sigma_{in}(k)$. Thus, a resonance can only result in peaks, not dips, in these cross sections.

2.7 The Scattering Matrix

We consider scattering from an initial state ψ_i (incoming waves only) to a final state ψ_f (outgoing waves only). Let ψ_i and ψ_f be vectors where the components are labeled by the quantum numbers j, T, parity, and channel number. A channel is specified by listing the numbers of each type of particle present. The scattering matrix S is defined:

$$\psi_f = S\psi_i \tag{2.49}$$

The rows and columns of S are labeled by total angular momentum j, total isotopic spin T, parity, and channel number. The S-matrix elements will be functions of the total energy E. The transformation (2.49) is always possible if we assume the superposition principle and that both the ψ_i and ψ_f are formed from a complete set of states. Since total probability must be conserved the S-matrix must be unitary,

$$\langle \psi_f | \psi_f \rangle = \langle S\psi_i | S\psi_i \rangle = \langle \psi_i | S^\dagger S | \psi_i \rangle = \langle \psi_i | \psi_i \rangle$$

Hence,

$$S^\dagger S = I \tag{2.50}$$

Since j, T, and parity are conserved, S will have off-diagonal elements connecting only states with the same j, T, and parity.

Consider the case where the initial state is a two-particle channel and where the energy is low enough so that no inelastic channels are possible. Then only elastic scattering can occur, and the S-matrix is diagonal. It will have the form,

$$S = \begin{pmatrix} e^{2i\delta_1} & 0 & 0 & \vdots \\ 0 & e^{2i\delta_2} & 0 & \vdots \\ 0 & 0 & e^{2i\delta_3} & \vdots \\ \cdots & \cdots & \cdots & \vdots \end{pmatrix} \tag{2.51}$$

where $\delta_1, \delta_2, \delta_3, \ldots$ are all real numbers. From the definition (2.49), then, the effect of elastic scattering is just to shift the phase of the initial state by 2δ. This is just the usual phase shift defined in sect. 2.1.

If there is more than one channel open in the final state, the S-matrix will no longer be diagonal. The phase shifts occurring in the diagonal elements will be complex. As before the imaginary part can be used to define an absorption parameter $\eta = e^{-2\,\mathrm{Im}\,\delta}$. The diagonal elements will then be of the form $\eta e^{2i\delta}$ where δ is real. From eq. (2.33) it is clear that

$$\sigma_{\mathrm{el}} = 2\pi\lambda\!\!\!^{-2}\ \Sigma\,(j + \tfrac{1}{2})|1 - S_{ii}|^2 \tag{2.52}$$

The S-matrix completely determines the scattering properties of a system of particles. Here we note that invariance under time reversal implies

$$S_{ij} = S_{ji} \tag{2.53}$$

provided that all spins are reversed or that there is no net polarization of any of the particles. Equation (2.53) is easily proved. Assume for the moment a system without spins. Let $\psi_a, a = 1, 2, \ldots, N$, be a complete set of *incoming* wave functions. The subscript a labels the channel and isotopic spin. The time reversed solution to ψ_a will be $\psi_a{}^*$. It will represent an *outgoing* wave. If the Hamiltonian describing the interaction is invariant under time reversal, then $\psi_a{}^*$ will be a valid final state. Thus, we can write

$$\psi_a^* = \sum_b S_{ba} \psi_b$$

Hence,

$$\psi_a = \sum_b S_{ba}^* \psi^*$$

Substituting we get,

$$\psi_a{}^* = \sum_b \sum_c S_{ba} S_{cb}{}^* \psi_c{}^*$$

Thus,

$$\sum_b S_{cb}{}^* S_{ba} = \delta_{ac}$$

Or,

$$SS^* = I$$

Taking the complex conjugate of (2.50) gives

$$S^{Tr} S^* = I$$

Comparing the last two matrix equations gives

$$S = S^{Tr}$$

This is just eq. (2.53). So far spins have been neglected. If they exist, relation (2.53) must be amended somewhat. We define the time reversed channel $-a$ to channel a as identical to a in all respects except that all spins are reversed. Then (2.53) can be written

$$S_{ij} = S_{-j, -i} \tag{2.53'}$$

If the S-matrix is diagonalized, the diagonal elements will again be of the form $e^{2i\delta}$ where the *eigen-phase* shifts δ are real. We shall see later that the same set of data can be fit by quite different sets of phase shifts and absorption parameters. The eigen-phase shifts, however, may be quite similar for all sets.

Consider the case where in addition to the elastic channel there is one inelastic channel in which scattering occurs for a given j, T, and parity. The S-matrix reduces to a 2×2 submatrix describing scattering in this state. We assume this submatrix is symmetric (time reversal invariance) and write,

$$S = \begin{pmatrix} S_{11} & S_{12} \\ S_{12} & S_{22} \end{pmatrix} \tag{2.54}$$

This matrix can be diagonalized by a real orthogonal matrix (Ref. 8)

$$\begin{pmatrix} e^{i\delta_a} & 0 \\ 0 & e^{i\delta_b} \end{pmatrix} = \begin{pmatrix} \cos\epsilon & -\sin\epsilon \\ \sin\epsilon & \cos\epsilon \end{pmatrix} \begin{pmatrix} S_{11} & S_{12} \\ S_{12} & S_{22} \end{pmatrix} \begin{pmatrix} \cos\epsilon & \sin\epsilon \\ -\sin\epsilon & \cos\epsilon \end{pmatrix} \tag{2.55}$$

where δ_a and δ_b are the eigen-phases in the two channels. From (2.55),

$$S_{11} = \eta_{11} e^{2i\delta_{11}} = \cos^2\epsilon e^{2i\delta_a} + \sin^2\epsilon e^{2i\delta_b} \tag{2.56}$$

where we have assumed that subscript 1 refers to the two-body channel. Hence,

$$\eta_{11} = \sqrt{1 - [\sin 2\epsilon \ \sin(\delta_a - \delta_b)]^2} \tag{2.57}$$

and

$$\cot 2\delta_{11} = \frac{\cos 2\delta_a \cos^2\epsilon + \cos 2\delta_b \sin^2\epsilon}{\sin 2\delta_a \cos^2\epsilon + \sin 2\delta_b \sin^2\epsilon} \tag{2.58}$$

We note that as $\eta_{11} \longrightarrow 0$, $\delta_a - \delta_b \longrightarrow 90$ deg. and $\epsilon \longrightarrow 45$ deg. independent of δ_{11}. If we assume that $\delta_b \cong 0$, then

$$\cot 2\delta_{11} \cong \cot 2\delta_a + \tan^2\epsilon \ \mathrm{cosec}\, 2\delta_a \tag{2.59}$$

If $\delta_a \longrightarrow 90°$ then $\delta_{11} \longrightarrow 0$ or 90 deg. independent of ϵ. If however $\delta_b \neq 0$, then by relation (2.58) $\delta_a = 90$ deg. does not imply $\delta_{11} = 0$ or 90 deg.

From (2.57) we see that $\eta_{11} = 1$ when $\epsilon = 0$ or $\delta_a = \delta_b$. This is just the case of pure elastic scattering. It is easy to show from eq. (2.55) that the S-matrix element $S_{12} = 0$ when $\epsilon = 0$ or $\delta_a = \delta_b$. Thus, the S-matrix becomes diagonal as expected.

2.8 Resonances

There are many ways of defining a resonance. All are more or less equivalent. Following the S-matrix theorists, we will define it simply as a pole of order one in the scattering amplitude for a single partial wave at a complex energy. Near the resonant energy, then, the resonant partial-wave amplitude will be dominated by this pole and will be given approximately by

$$\frac{\eta_{\ell j}(E) \exp \{2i\delta_{\ell j}(E)\} - 1}{2i} = a_{\ell j}(E) \propto \frac{1}{E_r - i\dfrac{\Gamma}{2} - E} \tag{2.60}$$

Here E is the total energy in the barycentric system, and E_r is the resonance energy. From (2.33) and (2.41), Γ (positive) will be the full width at half maximum in the integrated cross sections $\sigma_{e1}(e)$ and $\sigma_{tot}(E)$. Then $\sigma_{in}(E) = \sigma_{tot}(E) - \sigma_{e1}(E)$ will have the same width.

In order that the expression (2.60) dominate the cross section as E approaches E_r along the real axis, it is necessary that Γ be not too large. That is, the pole approximation for $a_{\ell j}(E)$ will only be valid for a sharp resonance when E is near E_r. It is possible to make (2.60) more general by allowing Γ to be a function of E.

Alternatively, we might say a resonance occurs whenever a phase shift goes rapidly through 90 deg. This definition is not as general as the pole definition. We will see that the pole definition allows for resonances in which a phase shift goes through 0 deg. also. We can even more generally say that a resonance occurs whenever an *eigen-phase* goes through 90 deg. This is equivalent to saying that the *eigen*-partial-wave amplitude satisfies (2.60). We saw in the last section that in the two channel approximation the phase shift could go through 0 or 90 deg. when the *eigen-phase* went through 90 deg. It could also do neither. The conditions for this will be made clear in this section.

We now define Γ_{el}, positive, such that (2.60) becomes,

$$a_{\ell j}(E) = \frac{\frac{1}{2}\Gamma_{el}}{E_r - i\dfrac{\Gamma}{2} - E} \tag{2.61}$$

The signs of the various terms are chosen so that at a resonance where the phase shift goes through 90 deg., it *increases* through 90 deg. rather than decreases. The reason for this will be seen later. If we assume that this term dominates the scattering amplitude near the resonance energy, then we can write from (2.33) and (2.41)

$$\sigma_{el}(E) = \pi \lambda\!\!\!^{-2}(j + \tfrac{1}{2}) \frac{\Gamma_{el}^2}{(E - E_r)^2 + \dfrac{\Gamma^2}{4}} \tag{2.62}$$

$$\sigma_{tot}(E) = \pi \lambda\!\!\!^{-2}(j + \tfrac{1}{2}) \frac{\Gamma_{el}\Gamma}{(E - E_r)^2 + \dfrac{\Gamma^2}{4}} \tag{2.63}$$

$$\sigma_{in}(E) = \pi \lambda\!\!\!^{-2}(j + \tfrac{1}{2}) \frac{\Gamma_{el}\Gamma_{in}}{(E - E_r)^2 + \dfrac{\Gamma^2}{4}} \tag{2.64}$$

where

$$\Gamma_{in} = \Gamma - \Gamma_{el} \tag{2.65}$$

Equations (2.62), (2.63), and (2.64) are called the Breit-Wigner one-level resonance formulas [Ref. 7]. The Γ_{el}, Γ_{in}, and Γ are called the elastic, inelastic, and total widths respectively.

It is clear that the maxima in $\sigma_{el}(E)$, $\sigma_{in}(E)$, $\sigma_{tot}(E)$ all occur when $E = E_r$ provided the Γ's are independent of E. From (2.61), $a_{\ell j}(E)$ is purely imaginary at the resonance energy. Thus, $\eta_{\ell j}(E_r) \exp\{2i\delta_{\ell j}(E_r)\}$ must be real. This means that $\delta_{\ell j}(E) = 0$ or $\pi/2$. From (2.62) and (2.63)

$$2\,\frac{\sigma_{el}(E_r)}{\sigma_{tot}(E_r)} = \frac{\Gamma_{el}}{\Gamma} \tag{2.66}$$

From (2.61), at $E = E_r$

$$\eta_{\ell j}(E_r) \exp\{2i\delta_{\ell j}(E_r)\} = 1 - 2\,\frac{\sigma_{el}(E_r)}{\sigma_{tot}(E_r)} = 1 - 2\,\frac{\Gamma_{el}}{\Gamma} \tag{2.67}$$

There are two cases:

Case 1: $\quad 1/2 < \Gamma_{el}/\Gamma \leqslant 1, \quad \delta_{\ell j}(E_r) = \pm\pi/2$

and

$$\eta_{\ell j}(E_r) = 2\,(\Gamma_{el}/\Gamma) - 1 \tag{2.68}$$

Case 2: $\quad 0 \leqslant \Gamma_{el}/\Gamma < 1/2, \quad \delta_{\ell j}(E_r) = 0, \pm\pi$

and

$$\eta_{\ell j}(E_r) = 1 - 2\,(\Gamma_{el}/\Gamma) \tag{2.69}$$

Case 1 corresponds to a resonance with the resonant phase shift passing through 90 deg. This includes the pure elastic resonance where

$$\Gamma_{el}/\Gamma = 1 \text{ and } \eta_{\ell j}(E_r) = 1 \tag{2.70}$$

Case 2 corresponds to a resonance with the resonant phase shift passing through 0 deg. This is a highly inelastic resonance:

$$\frac{\sigma_{el}(E_r)}{\sigma_{tot}(E_r)} < \frac{1}{2} \tag{2.71}$$

or

$$\frac{\sigma_{el}(E_r)}{\sigma_{in}(E_r)} < 1 \tag{2.72}$$

If $\Gamma_{el}/\Gamma = 1/2$, $\delta_{\ell j}(E_r)$ is indeterminate.

As can be seen from (2.62), (2.63), and (2.64), the maximum values $\sigma_{el}(E_r)$, $\sigma_{tot}(E_R)$, and $\sigma_{in}(E_r)$ decrease as the resonance becomes more and more inelastic, that is, as $(\Gamma_{el}/\Gamma) \to 0$.

Define:

$$x = \frac{\Gamma_{el}}{\Gamma} = \frac{\sigma_{el}(E)}{\sigma_{tot}(E)} \tag{2.73}$$

and

$$\epsilon = \frac{E_r - E}{\Gamma} \tag{2.74}$$

From (2.61) we can solve for η and $\sin 2\delta$ as functions of x and ϵ. The results are:

$$\eta = \left[1 - \frac{x(1-x)}{\epsilon^2 + \frac{1}{4}}\right]^{1/2} \tag{2.75}$$

$$\sin 2\delta = \frac{\epsilon x}{[(\epsilon^2 + \frac{1}{4})\{\epsilon^2 + (x - \frac{1}{2})^2\}]^{1/2}} \tag{2.76}$$

For an elastic resonance where $x = 1$, we have the simple results

$$\eta = 1, \quad \sin 2\delta = \frac{\epsilon}{\sqrt{\epsilon^2 + \frac{1}{4}}} \tag{2.77}$$

Let us assume that x is not a function of energy. This means the *ratios* of the three cross sections $\sigma_{el}(E)$, $\sigma_{tot}(E)$, and $\sigma_{in}(E)$ are independent of energy. We can then plot η and δ versus ϵ for various values of x. This is shown in Fig. 2.1.

2.9 The Wigner Condition

Under the assumption that the radius of interaction is finite, there is an important constraint on the rate of change of a phase shift [Ref. 9]. Consider a wave packet resolved into its incoming and outgoing spherical waves. We approximate a wave packet as the superposition of two waves of slightly different frequency and wave number. Thus,

$$\psi_{in} = \frac{1}{r}\,[e^{-i(kr+\omega t)} + e^{-i[(k+\Delta k)r+(\omega+\Delta\omega)t]}]$$

By eq. (2.7) there will be a relative phase difference between the incoming and outgoing waves of 2δ. Thus,

$$\psi_{out} = \frac{1}{r}\,[e^{-i[kr-\omega t+2\delta]} + e^{i[(k+\Delta k)r-(\omega+\Delta\omega)t+2(\delta+\Delta\delta)]}]$$

The two incoming waves will be in phase if

$$r\Delta k - t\Delta\omega = 0$$

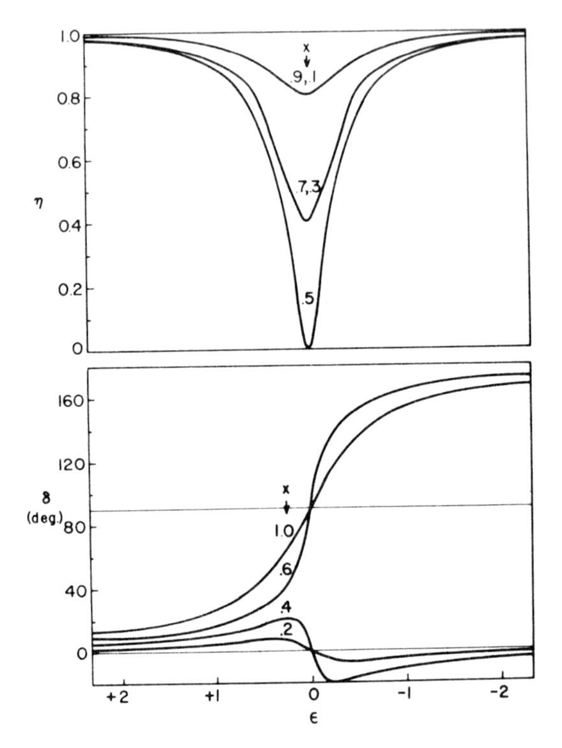

FIG. 2.1 Plot of eqs. (2.75) and (2.76).

Thus, $r_i = \left(\dfrac{\Delta\omega}{\Delta k}\right) t$ defines the center of the incoming wave packet. It moves with group velocity $\Delta\omega/\Delta k$.

Similarly the outgoing waves will be in phase if

$$r\Delta k - t\Delta\omega + 2\Delta\delta = 0$$

Hence, $r_0 = -2\dfrac{\Delta\delta}{\Delta k} + \left(\dfrac{\Delta\omega}{\Delta k}\right) t = -2\dfrac{\Delta\delta}{\Delta k} + r_i$ defines the center of the outgoing wave packet.

Let R be the finite radius of interaction. If the outgoing wave is not to leave the scattering region before the arrival of the incoming wave on the surface R, then for $r_i = R$ we must have $r_0 < R$. This yields (taking the limit $\Delta \longrightarrow 0$) the inequality

$$d\delta/dk > -R \tag{2.78}$$

In practice, this relation is useful only in the region of a resonance where some phase shift is rapidly changing. In this case, it will be possible to satisfy (2.78) only if the phase shift is increasing rather than decreasing. This means that for an elastic resonance the phase shift must increase through 90 deg. at the peak of the resonance. For a resonant amplitude as defined in (2.61), $d\delta/dk$ is positive. It is in fact to guarantee that (2.78) will be satisfied that we have, as previously mentioned, chosen the signs in (2.61).

2.10 Isotopic Spin

To this point nothing has been said about the various charge states possible in π-N scattering. Only spin angular momentum has been taken into account. By the hypothesis of charge independence, as explained in Chapter 1, the scattering amplitude must be a linear combination of a $T = \frac{1}{2}$ and a $T = \frac{3}{2}$ scattering amplitude. Thus, the $a_{\ell j}$ defined in (2.6) and (2.44) can be written for each of the 10 reactions (1.12) as,

$$a_{\ell j} = \gamma_1 a_{\ell j}(T = \tfrac{1}{2}) + \gamma_3 a_{\ell j}(T = \tfrac{3}{2}) \tag{2.79}$$

The coefficients can be simply derived from (1.12) and (1.15). They are given in Table 2.1.

TABLE 2.1

Reaction	γ_1	γ_3
$\pi^+ p \longrightarrow \pi^+ p$	0	1
$\pi^- p \longrightarrow \pi^- p$	2/3	1/3
$\pi^- p \longrightarrow \pi^0 n$	$-\sqrt{2}/3$	$\sqrt{2}/3$
$\pi^+ n \longrightarrow \pi^+ n$	2/3	1/3
$\pi^+ n \longrightarrow \pi^0 p$	$-\sqrt{2}/3$	$\sqrt{2}/3$
$\pi^- n \longrightarrow \pi^- n$	0	1
$\pi^0 p \longrightarrow \pi^0 p$	1/3	2/3
$\pi^0 n \longrightarrow \pi^0 n$	1/3	2/3
$\pi^0 n \longrightarrow \pi^- p$	$-\sqrt{2}/3$	$\sqrt{2}/3$
$\pi^0 p \longrightarrow \pi^+ n$	$-\sqrt{2}/3$	$\sqrt{2}/3$

Using (2.79) we gain an important constraint on the charge exchange differential cross section (1.13) (iii) at 0 and 180 deg. From the table we have,

$$\begin{aligned} f_0(\theta,k) &= \sqrt{\tfrac{2}{3}} f_3(\theta,k) - \sqrt{\tfrac{2}{3}} f_1(\theta,k) \\ &= \sqrt{2}\,[f_3(\theta,k) - \{\tfrac{2}{3} f_3(\theta,k) + \tfrac{1}{3} f_1(\theta,k)\}] \\ &= \sqrt{2}\,[f_+(\theta,k) - f_-(\theta,k)] \end{aligned} \tag{2.80}$$

where $f_0(\theta,k)$, $f_\pm(\theta,k)$, $f_1(\theta\ k)$, $f_3(\theta,k)$ are the non-spin flip amplitudes defined in (2.20) for charge exchange, $\pi^\pm p$, $T = \frac{1}{2}$ scattering and $T = \frac{3}{2}$ scattering, respectively. At 0 and 180 deg., $P_\ell^1(\cos\theta) = 0$ so that the spin-flip amplitude defined in (2.20) $g(\theta,k) = 0$. Thus,

$$\frac{d\sigma^0}{d\Omega} = 2|f_+ - f_-|^2 \geqslant 2[|f_+| - |f_-|]^2 \tag{2.81}$$

at 0 and 180 deg. where $d\sigma^0/d\Omega$ is the charge exchange differential cross section. Hence,

$$\frac{d\sigma^0}{d\Omega} \geqslant 2\left|\sqrt{\frac{d\sigma^+}{d\Omega}} - \sqrt{\frac{d\sigma^-}{d\Omega}}\right|^2 \text{ at 0 and 180 deg.} \tag{2.82}$$

The experimental data must satisfy this relation if charge independence is valid.

REFERENCES

1. M. L. Goldberger and K. M. Watson, *Collision Theory*, John Wiley and Sons (1964).
2. For a complete discussion see any standard text of quantum mechanics such as L. I. Schiff, *Quantum Mechanics*, McGraw Hill Book Co., (1955).
3. For proof of this assertion see Ch. 8, W. S. C. Williams, *An Introduction to Elementary Particles*, Academic Press, (1961).
4. R. A. Sorensen, *Phys. Rev. 112*, 1813 (1958).
5. F. T. Solmitz, *Phys. Rev. 94*, 1799 (1954).
6. L. D. Roper, R. M. Wright, and B. T. Feld, *Phys. Rev. 138*, B190 (1965).
7. G. Breit and E. P. Wigner, *Phys. Rev. 49*, 519 (1936).
8. H. Goldberg, Phys. Rev. *151*, 1186 (1966).
9. E. P. Wigner, *Phys. Rev. 98*, 145 (1955).

CHAPTER 3

Total Cross Section

In this chapter the π-N total cross sections will be discussed to see what information can be gained from them.

3.1 Experimental Method

The total cross section is determined by measuring the attenuation of a π beam resulting from passage through a target of hydrogen. Counters are often used. The pions are produced by protons from an accelerator passing through a target, usually a metal such as aluminum, copper, or uranium. Pions of the desired momentum are selected and focussed by a series of bending and quadrupole magnets.

A typical experimental setup is shown in Fig. 3.1. The incident particles are recorded by three counters S_1, S_2, and S_3. Four more counters $T_1, \ldots, T_4$

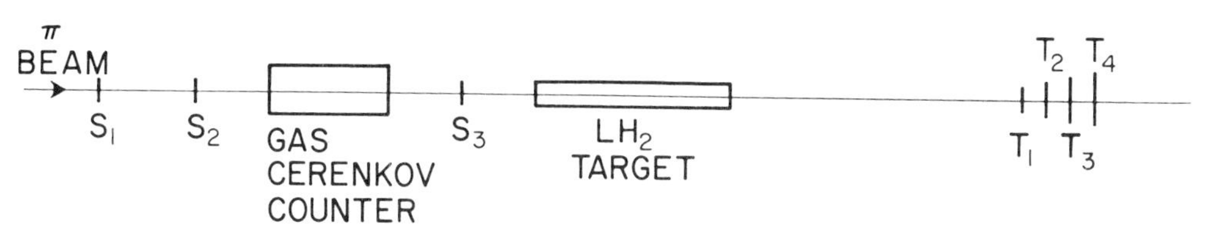

FIG. 3.1 Typical counter set-up to measure $\pi^{\pm}-p$ total cross sections.

record (in coincidence with the S-counters) those particles remaining in the beam after passage through the liquid hydrogen target. The attenuation due to the target is given by

$$N = N_0 e^{-\sigma x} \tag{3.1}$$

where N_0 = the flux of incident particles,
N = the flux of particles remaining after passage through the target,
σ = the total cross section in cm^2,
x = the thickness of the hydrogen target in protons/cm^2.

Thus,

$$\sigma = \frac{1}{x} \ln(N_0/N) \tag{3.2}$$

In practice the T-counters must subtend a finite solid angle. Thus, particles can scatter through a small angle and still pass through the T-counters. To correct for this, data are taken with the T-counters moved successively further and further from the target. At each distance (and solid angle) a cross section is determined. This cross section is plotted as a function of the solid angle subtended by the T-counters. The plot is then extrapolated to zero solid angle. An idealized curve is shown in Fig. 3.2. The dotted line is the extrapolation to zero solid angle. The measured curve rises sharply for very small solid angles. This is caused by Coulomb scattering, which is sharply peaked at zero scattering angle.

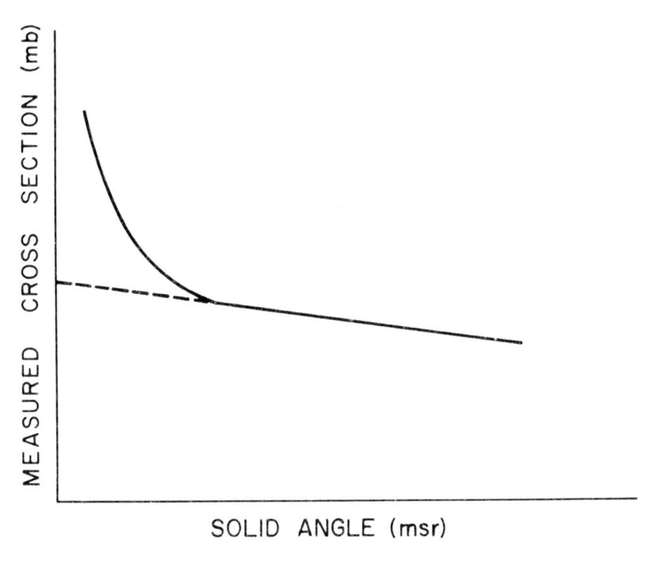

FIG. 3.2 Idealized curve of measured total cross section versus the solid angle subtended by the T-counters of Fig. 3.1.

The Cerenkov counter in Fig. 3.1 is used to determine the inevitable electron and muon contamination in the beam. At high energies a gas counter is used. The pressure is varied. The threshold pressure is determined by

$$n = 1/\beta \tag{3.3}$$

where n = index of refraction of the gas and βc = the velocity of the particle. Since the bending magnets in the beam ensure that all the particles have nearly the same momentum, electrons, muons, and pions will have different thresholds. A typical pressure curve is shown in Fig. 3.3. The relative heights of the plateaus give the fraction of electrons, muons, and pions in the beam. The electrons and muons will experience only Coulomb scattering in the hydrogen target. Thus, they will always be counted by the T-counters. If their relative numbers in the incoming beam are known they can be subtracted.

A π^+ beam will generally have many times more protons than pions. Protons will be scattered by the hydrogen target. They cannot be subtracted out in the simple way that electrons and muons can. A velocity selector is sometimes used to eliminate most of them. At low momenta protons can be eliminated by stopping them in an absorber. Also, they can be eliminated by their

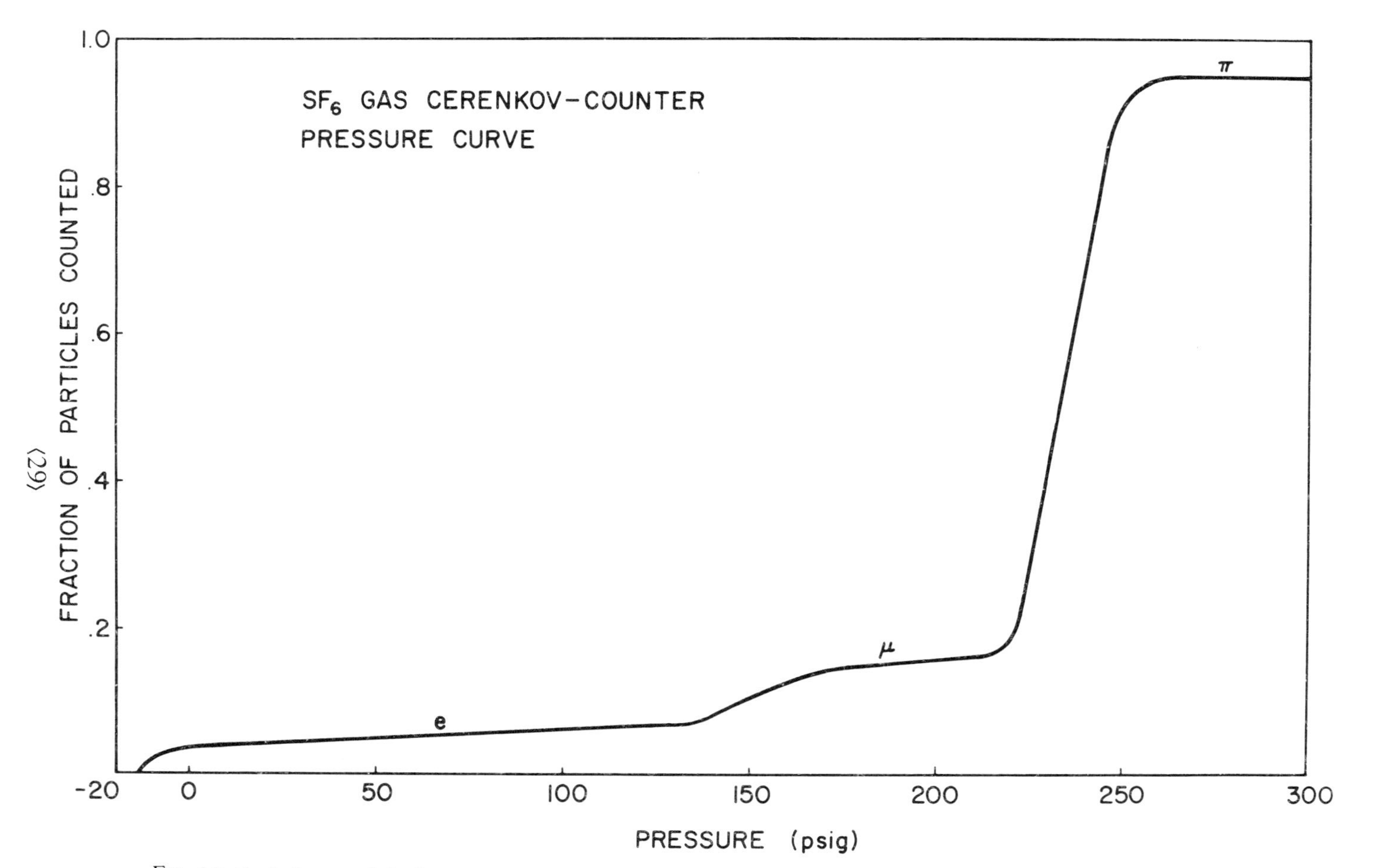

FIG. 3.3 Typical curve of the fraction of beam particles counted versus pressure for the Cêrenkov counter shown in Fig. 3.1.

different flight times through counters S_1, S_2, and S_3. For this purpose these counters are separated in space as much as possible.

3.2 Experimental Results

In Figs. 3.4 and 3.5 are shown the total cross sections for $\pi^{\pm}p$ and $\pi^{-}p$ from 0 to 20 GeV [Ref. 1]. Also included are the integrated elastic and charge exchange cross sections. Fig. 3.6 shows the $T = \frac{1}{2}$ and $T = \frac{3}{2}$ cross sections deduced from very accurate measurements made between 2 and 7 BeV [Ref. 2].

The most striking feature of the total cross sections is the many peaks that appear. In Table 3.1 are listed all the experimentally known bumps (as illustrated in Figs. 3.4, 3.5, and 3.6), along with their widths, spins, isotopic spins, and masses, which are defined as the total energy in the barycentric system. The determination of the properties of the "resonances" as listed in Table 3.1 constitutes the major problem of π-N scattering experiments.

TABLE 3.1
List of π-N Resonances

Lab energy	Mass	Full width	Isotopic spin	Spin
(MeV)	(MeV)	(MeV)		
0	938	0	$\frac{1}{2}$	$\frac{1}{2}$
200	1,238	90	$\frac{3}{2}$	$\frac{3}{2}$
611	1,519	110	$\frac{1}{2}$	$\frac{3}{2}$
870	1,670	200	$\frac{3}{2}$	$\frac{1}{2}$
900	1,688	120	$\frac{1}{2}$	$\frac{5}{2}$
1,350	1,920	260	$\frac{3}{2}$	$\frac{7}{2}$
1,970	2,190	200	$\frac{1}{2}$	?
2,510	2,420	310	$\frac{3}{2}$	?
3,100	2,650	360	$\frac{1}{2}$	?
3,630	2,850	400	$\frac{3}{2}$	?

We speak loosely of these peaks as resonances or particles. The traditional distinction between the two has been that a particle has a lifetime long enough to leave the center of interaction before decaying, whereas a resonance has not. This is a somewhat artificial distinction. The lifetime of a particle is determined by what it can decay into and whether or not the possible decays conserve various quantities which are known to have invariance properties. That is, a short or long lifetime is merely an accident of the spectrum of particles. Thus, for example, the nucleon is a resonance which just happens to be stable because there is nothing into which it can decay, conserving energy,

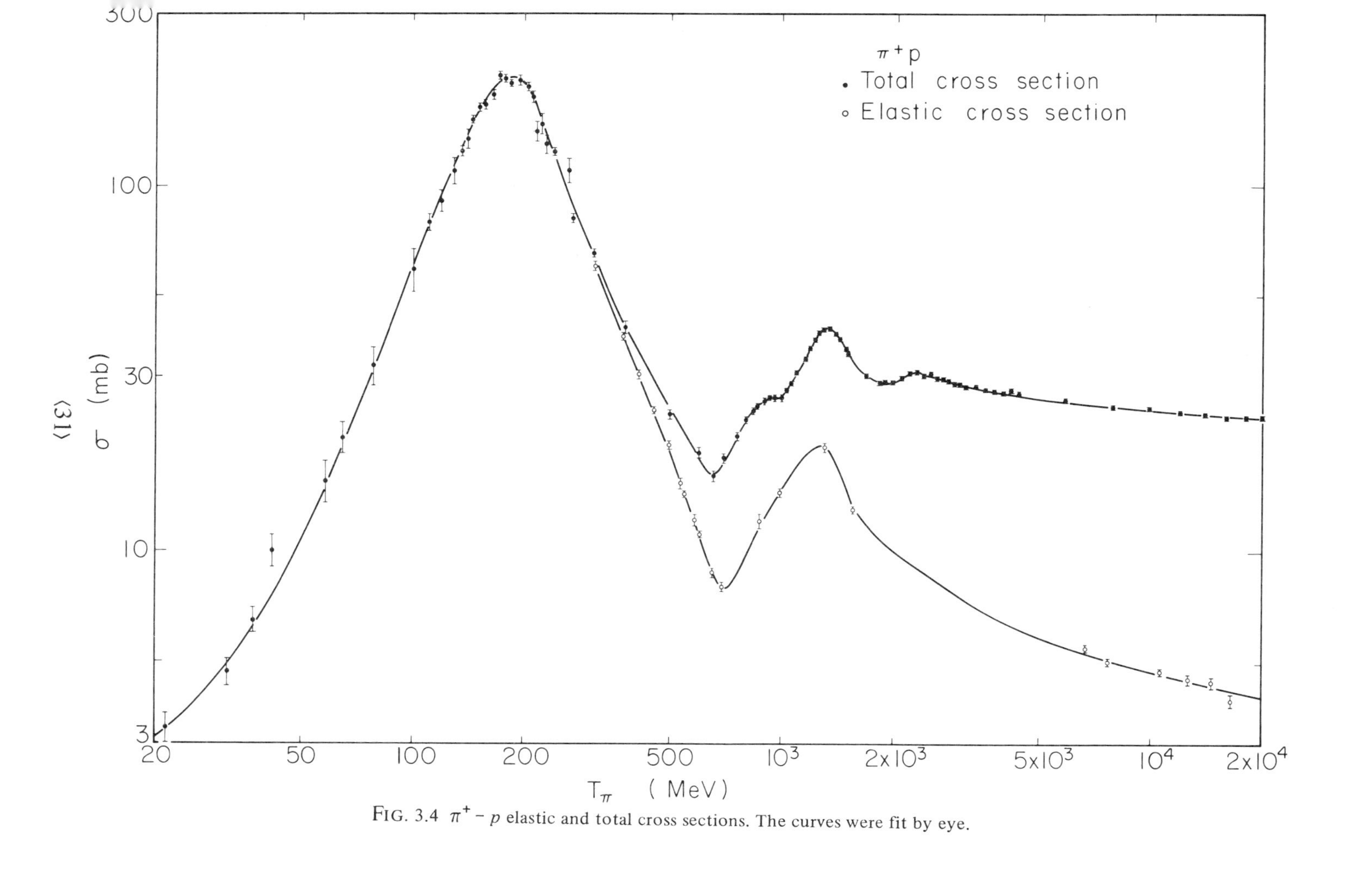

FIG. 3.4 $\pi^+ - p$ elastic and total cross sections. The curves were fit by eye.

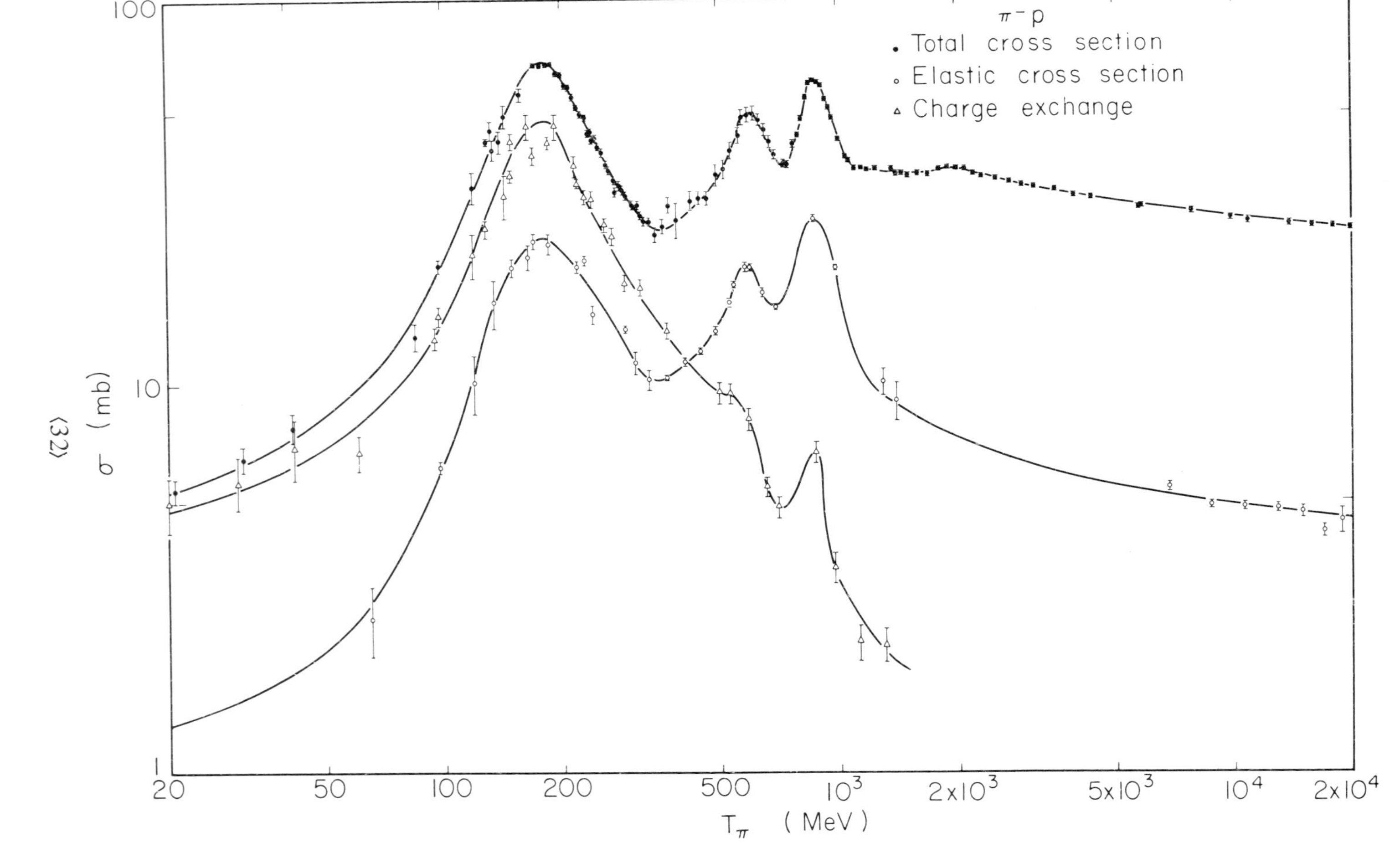

FIG. 3.5 $\pi^- - p$ elastic, charge exchange and total cross sections. The curves were fit by eye.

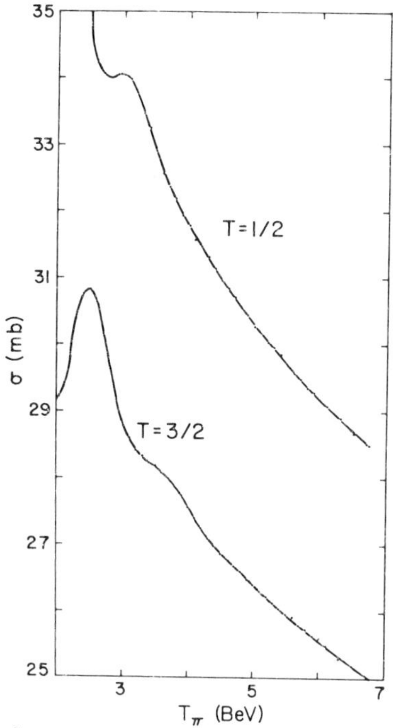

FIG. 3.6 $T = \frac{1}{2}$ and $T = \frac{3}{2}$ total cross sections between 2 and 7 BeV. The curves were fit by eye.

angular momentum, and baryon number. Thus, it is included in the list of resonances in Table 3.1.

A feature of these resonances that one notices from Table 3.1 is that they are all quite narrow. With the exception of the nucleon, which is stable, the full width of all the resonances is of the order of 10 per cent of the mass. This makes it plausible that these resonances are each due to a small number of angular momentum states. It would seem unlikely that two different phase shifts would pass rapidly through 90 deg. at precisely the same energy. Thus, these resonances appear to be the same as the traditional one-level resonances so often found in nuclear physics. We shall see later, however, that the results of phase shift analysis appear to disagree with the above argument. Nevertheless, for the moment we will accept it.

From eq. (2.40) we note that the total cross section is *linear* in the imaginary parts of the partial wave amplitudes. Thus, there can be no interference between them. If we have one-level resonances, then the height of a resonance is determined by (2.47),

$$\sigma_{\text{tot}}^j \leqslant \pi \lambda\!\!\!^{-2} (2j + 1)(\eta_j + 1) \tag{3.4}$$

This height must be measured from the non-resonant part of the total cross section. In the region of the 200 MeV resonance, this is almost 0. At higher energies, however, there is considerable non-resonant background; and for nearby resonances which overlap, such as those at 600 and 900 MeV, it takes considerable care to determine just what is the non-resonant part of the cross section.

3.3 The 200 MeV Resonance

The oldest known and best understood πp resonance is the one at 200 MeV. We can see from Figs. 3.4 and 3.5 that the ratio of $\sigma(\pi^+ p \longrightarrow \pi^+ p)$: $\sigma(\pi^- p \longrightarrow \pi^\circ n)$: $\sigma(\pi^- p \longrightarrow \pi^- p)$ at 200 MeV is very nearly 9:2:1. A glance at eqs. (1.16) shows that exactly this ratio is predicted if $F_1 = 0$. This agreement with experiment indicates two things. First, the assumption that all the scattering amplitudes depend on only two independent amplitudes is valid. That is, the hypothesis of charge independence is correct. Second, the 200 MeV resonance is a resonance in the state $T = \frac{3}{2}$.

At the peak of the resonance the total cross section for $\pi^+ p$ is very closely equal to $8\pi\lambda\!\!\!^{-2}$. This is the value predicted by (3.4) if $j = \frac{3}{2}$ and $\eta_j = 1$. The inelastic cross section is expected to be small, since we are very near threshold for single pion production, which is the first inelastic channel. This is eminently borne out by experiment as shown in Fig. 3.1. The total cross section alone, then, allows the conclusion that the resonant angular momentum is $j = \frac{3}{2}$. At this point, however, it is not known whether it is a $P\frac{3}{2}$, or a $D\frac{3}{2}$ resonance. We can, however, hazard a guess using Table 1.1. Since $R/\lambda\!\!\!^{-}_B < 2$ at 200 MeV, $P\frac{3}{2}$ is clearly favored. This will be verified later by the phase shifts deduced from the differential cross sections. We refer to this as a P_{33} resonance where the subscripts are 2T and $2j$ respectively.

3.4 The Higher Resonances

Accepting now the hypothesis of charge independence, the isospin of the higher resonances can be deduced. From eq. (1.24) we can deduce the total cross section in isotopic spin state $T = \frac{1}{2}$. The $T = \frac{3}{2}$ cross section is, of course, the same as the $\pi^+ p$ total cross section. From eqs. (1.24) and (1.25) those bumps that appear in both $\pi^+ p$ and $\pi^- p$ total cross sections are due to a $T = \frac{3}{2}$ interaction. Those that appear only in the $\pi^- p$ cross section are due to a $T = \frac{1}{2}$ interaction.

Table 3.1 exhibits an interesting pattern in the isotopic spins of the various resonances. For every $T = \frac{1}{2}$ resonance (including the nucleon), there is a $T = \frac{3}{2}$ resonance at an energy a few hundred MeV higher. These isotopic pairs of resonances seem to be separated by several hundred MeV.

The last column of Table 3.1 gives the total angular momenta believed to be responsible for the various resonances. The evidence for these assignments will be given in later chapters. For the resonances above 200 MeV, it is not possible to make a reasonable estimate of the angular momentum involved by comparing the total cross section with $\pi\lambda\!\!\!^{-2}$. This is because, first, the inelasticity parameter $\eta_{\ell j}$ defined in (2.44) is not known. This can change the predicted resonant cross section by as much as a factor of 2. Second, as one can see from Figs. 3.4 and 3.5 the total cross section has a large non-resonant background at all the higher resonances. Thus, it is not possible to make an unambiguous subtraction. For these reasons, no experimental information about the angular momenta responsible for the higher resonances is obtainable from the total cross section.

3.5 The Pomeranchuk Theorem

This theorem relates the π^+p and π^-p total cross sections in the limit of infinite energy. It in fact says they are constant and equal [Ref. 3].

If the heights of the resonances in Fig. 3.1 are examined, we notice that they fall rapidly as the energy increases. These heights, measured above the non-resonant background, fall by about one order of magnitude moving from one isotopic pair to the next in the direction of increasing energy. By the time the isotopic pair at 3.1 and 3.63 GeV is reached, the height is only a few tenths mb, barely visible above the non-resonant background. This is undoubtedly due to the fact that, as the energy increases, more and more partial waves contribute to the cross section. A resonance in any one partial wave then has less and less overall effect. The point we wish to make here is simply to note the general flattening of the total cross section at very high energies. It becomes featureless.

In fact, as Pomeranchuk has argued, the total cross sections should become constant as $E \longrightarrow \infty$ [Ref. 3]. This follows simply from the fact that the π-N interaction has a finite range R. This implies a finite total cross section which should approach πR^2 as $E \longrightarrow \infty$.

As can be seen from Fig. 3.3, the charge exchange cross section appears to be approaching zero more rapidly than $\sigma_{el}^{\pm}(E)$ as the energy increases. This is to be expected, since charge exchange must compete with more and more inelastic channels as the energy increases while the total cross section remains constant.

From eqs. 1.16 we see that if the charge exchange cross section is much smaller than $\sigma_{el}^{\pm}(E)$, then

$$F_3 - F_1 << F_3, F_1$$

and hence

$$F_3 \cong F_1 \tag{3.5}$$

Then by eqs. (1.21), (1.23), and the optical theorem, eq. (2.40), the $\pi^{\pm}p$ total cross sections must be *equal* in the limit of infinite energy.

$$\sigma_{tot}^{+}(\infty) = \sigma_{tot}^{-}(\infty) \tag{3.6}$$

Also, by eqs. (1.16) the elastic differential cross sections for π^+p and π^-p become equal.

Figs. 3.4 and 3.5 show that the $\pi^{\pm}p$ total cross sections are indeed approaching each other at high energies. Intuitively, this is not surprising. As the energy increases, more and more channels open up for scattering. Thus, conservation of isotopic spin will not impose any limitation on either cross section. It should be emphasized, however, that there is no way to tell how rapidly the $\pi^{\pm}p$ cross sections should approach each other as $E \longrightarrow \infty$.

In the next chapter the Pomeranchuk theorem, eq. (3.6), will be shown to follow from dispersion theory.

REFERENCES

1. The data for Figs. 3.4 and 3.5 come from the following sources:

Total cross section

W. Galbraith, E. W. Jenkins, T. F. Kycia, B. A. Leontic, R. H. Phillips, and A. L. Read, *Phys. Rev. 138*, B913 (1965).

A. N. Diddens, E. W. Jenkins, T. F. Kycia, and K. F. Riley, *Phys. Rev.* Letters *10*, 262 (1963).

T. J. Devlin, J. Solomon, and G. Bertsch, *Phys. Rev.* Letters *14,* 1031 (1965).

Elastic cross section

P. M. Ogden, D. E. Hagge, J. A. Helland, M. Banner, J. F. Detoeuf, and J. Teiger, *Phys. Rev. 137,* B1115 (1965).

J. A. Helland, C. D. Wood, T. J. Devlin, D. E. Hagge, M. J. Longo, B. J. Moyer, and V. Perez-Mendez, *Phys. Rev. 134,* 1062 and 1079 (1964).

K. J. Foley, S. J. Lindenbaum, W. A. Love, S. Ozaki, J. J. Russell, and L. C. L. Yuan, *Phys. Rev.* Letters *11*, 425 (1963).

Charge exchange

C. B. Chiu, R. D. Eandi, A. C. Helmholz, R. W. Kenney, B. J. Moyer, J. A. Poirier, W. B. Richards, R. J. Cence, V. Z. Peterson, N. K. Sehgal, and V. J. Stenger, *Phys. Rev. 156,* 1415 (1967).

Selected data published prior to 1961 were also used. This has been compiled by V. S. Barashenkov and V. M. Maltsev, *Fortschritte der Phys. 9,* 549 (1961).

2. A. Citron, W. Galbraith, T. F. Kycia, B. A. Leontic, R. H. Phillips, A. Rousset, and P. H. Sharp, *Phys. Rev. 144,* 1101 (1966).
3. L. B. Okun and I. Ia. Pomeranchuk, J. *Exptl. Theoret. Phys.* (U.S.S.R.) *30,* 424 (1956) [Soviet Phys. J.E.T.P. *3*, 307 (1956)].

CHAPTER 4

The Forward Amplitude

4.1 Dispersion Relations

A relation between the real and imaginary parts of the forward scattering amplitude for the scattering of light on atoms has been known for some time. The relation, known as the Kramers-Kronig dispersion relation, is derived from the condition that the scattered wave should have zero amplitude until the incident wave reaches the scatterer [Ref. 1].

Kramers and Kronig showed that the requirement that light shall not propagate with speeds exceeding c leads to a relation between the real and imaginary parts of the index of refraction. The relation is,

$$\operatorname{Re} n(\omega) = 1 + \frac{2}{\pi} P \int_0^\infty d\omega' \, \frac{\omega' \operatorname{Im} n(\omega')}{\omega'^2 - \omega^2} \tag{4.1}$$

where $n(\omega)$ = the index of refraction, ω = the frequency of the light in the lab. The P indicates that the principal value of the integral is to be taken at the singularity $\omega' = \omega$. In addition to the causality assumption two further assumptions are necessary, namely,

$$\begin{aligned} n^*(-\omega) &= n(\omega) \\ \text{and} \quad \lim_{|\omega| \to \infty} n(\omega) &= 1 \end{aligned} \tag{4.2}$$

We will show that eq. (4.1) follows if eqs. (4.2) are assumed and if $n(\omega)$ is an analytic function in the upper-half complex ω-plane.[1] If analyticity is assumed we can use Cauchy's theorem and write,

$$n(\omega) = \frac{1}{2\pi i} \oint \frac{n(\omega')}{\omega' - \omega} d\omega' \tag{4.3}$$

The contour C will be taken to be a semi-circle in the upper-half ω-plane connected by a line along the real axis as shown in Fig. 4.1. A small semi-circle must be drawn to include the point ω. The contour along the semi-circle expanded to infinity gives a contribution to the integral in eq. (4.3) of $\frac{1}{2} n(\infty) = \frac{1}{2}$ by the second of eqs. (4.2). The small semi-circle around ω

[1] The elementary classical model for an index of refraction is based on a collection of damped electronic oscillators and gives an index of refraction,

$$n(\omega) = 1 + \frac{2\pi N e^2}{m} \sum_k \frac{f_k}{\omega_k^2 - \omega^2 - i\nu_k \omega}$$

where ω_k is the resonant frequency of the kth type of oscillator, ν_k its damping constant, and f_k the number of such oscillators per atom. This index of refraction is clearly analytic in the upper-half complex ω-plane. Furthermore, $n^*(-\omega) = n(\omega)$.

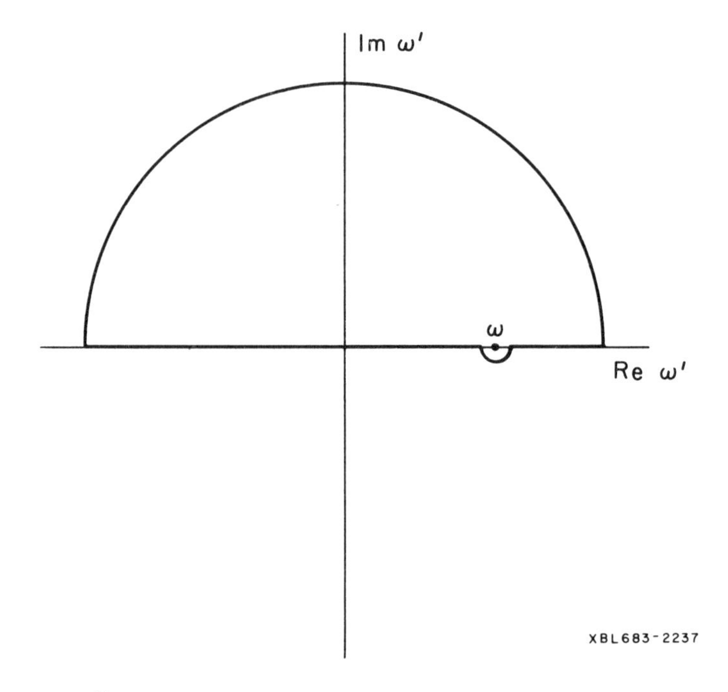

FIG. 4.1 Contour used in evaluating the integral in eq. 4.3.

gives a contribution $-\frac{1}{2}n(\omega)$. Thus, eq. (4.3) becomes

$$n(\omega) = 1 + \frac{1}{i\pi} P \int_{-\infty}^{\infty} \frac{n(\omega')}{\omega' - \omega} \, d\omega' \tag{4.4}$$

The integration in (4.4) is along the real axis in the ω-plane and P indicates the principal part of the integral is to be used. Taking the real part of eq. (4.4) and rearranging the integral over $\omega' < 0$ gives,

$$\operatorname{Re} n(\omega) = 1 + \frac{1}{\pi} P \int_{0}^{\infty} \left\{ \frac{\operatorname{Im} n(\omega')}{\omega' - \omega} - \frac{\operatorname{Im} n(-\omega')}{\omega' + \omega} \right\} d\omega' \tag{4.5}$$

The first of eqs. (4.2) reduces eq. (4.5) to eq. (4.1).

The macroscopic concept of index of refraction can be related to the scattering properties of individual atoms. This is done by means of the Lorentz relation,

$$n(\omega) = 1 + \frac{2\pi}{k^2} N f(\omega) \tag{4.6}$$

where $k = \omega/c$, the wave number of light of radial frequency ω, N is the number of scattering centers per cm^3, and $f(\omega)$ is the coherent forward scattering amplitude. Substituting eq. (4.6) into eq. (4.5) gives,

$$\operatorname{Re} f(\omega) = \frac{2k^2}{\pi} P \int_0^\infty \frac{d\omega'}{k'} \left\{ \frac{\operatorname{Im} f(\omega')}{\omega' - \omega} - \frac{\operatorname{Im} f(-\omega')}{\omega' + \omega} \right\} \tag{4.7}$$

Now eqs. (4.1) and (4.6) imply,

$$f(\omega) = f^*(-\omega) \tag{4.8}$$

Putting this last relation into eq. (4.7) gives,

$$\operatorname{Re} f(\omega) = \frac{2k^2}{\pi} P \int_0^\infty \frac{\omega'}{k'^2} \frac{\operatorname{Im} f(\omega')}{(\omega'^2 - \omega^2)} d\omega' \tag{4.9}$$

In eq. (4.6) it has been tacitly assumed that $f(0) = 0$. Otherwise, $n(\omega)$ would have a pole at $\omega = 0$. The expression on the right hand side of eq. (4.9) $\longrightarrow 0$ as $\omega \longrightarrow 0$ unless $n(\omega)$ has a pole at $\omega = 0$. If $f(0) \neq 0$, then eq. (4.9) must be modified to read,

$$\operatorname{Re} f(\omega) - \operatorname{Re} f(0) = \frac{2k^2}{\pi} P \int_0^\infty \frac{\omega'}{k'^2} \frac{\operatorname{Im} f(\omega')}{(\omega'^2 - \omega^2)} d\omega' \tag{4.10}$$

The total cross section $\sigma_{tot}(\omega)$ can be related to $\operatorname{Im} f(\omega)$ via the optical theorem. The proof given in Sect. (2.4) depended on conservation of total probability only. It did not depend on the mass and spin of the incident particles. For the scattering of light we can then write,

$$\operatorname{Im} f(\omega) = \frac{k}{4\pi} \sigma_{tot}(\omega) \tag{4.11}$$

Then eq. (4.10) can be written,

$$\operatorname{Re} f(\omega) - \operatorname{Re} f(0) = \frac{k^2}{2\pi^2} P \int_0^\infty \frac{\omega'}{k'} \frac{\sigma_{tot}(\omega')}{(\omega'^2 - \omega^2)} d\omega' \tag{4.12}$$

Equation (4.12) is generally referred to as the Kramers-Kronig dispersion relation.

It was suggested by Kronig that a causality condition might be extended to apply to the scattering of particles with mass [Ref.3]. This was first done by Karplus and Ruderman [Ref 4]. Rather than use a causality condition we will take a simpler route and indicate how these dispersion relations follow from an analyticity assumption. In our approach three assumptions are necessary: (a) $f(\omega)$, the forward scattering amplitude for particles of mass μ and total energy ω in the lab, is an analytic function of ω in the upper-half complex ω-plane, (b) relation (4.8) still holds, and (c) $(1/k) \operatorname{Im} f(\omega)$ is bounded as

$\omega \longrightarrow \infty$. Assumption (a) must be limited to the upper-half ω-plane since resonances will contribute poles in the lower half plane by eq. (2.61). Assumption (c) follows from the finite range of the π-N interaction and the optical theorem. By the optical theorem $\operatorname{Im} f(\mu) = 0$ since $k = 0$. It can in fact be shown that for $0 \leqslant \omega \leqslant \mu$, $\operatorname{Im} f(\omega) = 0$ provided there are no bound states [Ref.4]. Thus, the lower limit of the dispersion integral can be changed from 0 to μ. Analogous to eq. (4.12) we can write for particles of mass μ,

$$\operatorname{Re} f(\omega) - \operatorname{Re} f(\mu) = \frac{k^2}{2\pi^2} P \int_{\mu}^{\infty} \frac{\omega'}{k'} \frac{\sigma_{tot}(\omega')}{(\omega'^2 - \omega^2)} d\omega' \tag{4.13}$$

This dispersion relation can be correct only for the experimentally inaccessible reactions $\pi^{\circ} p \longrightarrow \pi^{\circ} p$ and $\pi^{\circ} n \longrightarrow \pi^{\circ} n$. For charged pions the dispersion relations become more complicated. Equation (4.8) is no longer correct. Since a negative (positive) energy $\pi^{\pm}$ behaves like a positive (negative) energy $\pi^{\mp}$ respectively, it is natural that eq. (4.8) be generalized for charged pions to read,

$$f_{\pm}(\omega) = f_{\mp}{}^*(-\omega) \tag{4.14}$$

where $f_{\pm}(\omega)$ is the coherent forward scattering amplitude for $\pi^{\pm} p \rightarrow \pi^{\pm} p$. Equation (4.13) would not satisfy eq. (4.14) if subscripts $\pm$ were simply added to the relevant quantities. If the linear combination $f_+(\omega) + f_-(\omega)$ is substituted into eq. (4.7) and eq. (4.14) used, the result is,

$$\operatorname{Re}\,[f_+(\omega) + f_-(\omega)] - \operatorname{Re}\,[f_+(\mu) + f_-(\mu)] =$$

$$\frac{2k^2}{\pi} P \int_0^{\infty} \frac{\omega'}{k'^2} \frac{\operatorname{Im} f_+(\omega') + \operatorname{Im} f_-(\omega')}{\omega'^2 - \omega^2} d\omega' \tag{4.15}$$

Equation (4.15) clearly satisfies eq. (4.14).

To determine $f_+(\omega)$ and $f_-(\omega)$ separately a dispersion relation for another linear combination is needed. If $f_+(\omega) - f_-(\omega)$ is substituted into eq. (4.7) and eq. (4.14) is used, the result is,

$$\operatorname{Re}\,[f_+(\omega) - f_-(\omega)] - \operatorname{Re}\,[f_+(\mu) - f_-(\mu)]\,\frac{\omega}{\mu} =$$

$$\frac{2\omega k^2}{\pi} P \int_0^{\infty} \frac{1}{k'^2} \frac{\operatorname{Im}\,[f+(\omega') - f_-(\omega')]}{\omega'^2 - \omega^2} d\omega' \tag{4.16}$$

Note that the second term on the left has been multiplied by ω/μ. This is to ensure that eq. (4.14) is satisfied. It will be shown in Sect. 4.3 that as $\omega \longrightarrow \infty$, $|f_{\pm}(\omega)| \leqslant a\omega$ where a is some constant. Thus, higher odd powers of ω/μ are not allowed to multiply the second term on the left in eq. (4.16). The left hand side of eq. (4.16) is still 0 at $\omega = \mu$ as required.

Equations (4.15) and (4.16) were rigorously derived by Goldberger [Ref. 5]. Instead of the analyticity assumption used here he used the equivalent in quantum mechanics of the causality assumption of Kramers and Kronig. The demand that waves do not propagate faster than the velocity of light is expressed by requiring that measurements of two observable quantities made at space-like points should not interfere. Thus, the causality condition is imposed by setting the commutator of two Heisenberg operators for the boson field equal to zero when these are taken at space-like points.

It will be noticed that the lower limits of the integrals in eqs. (4.15) and (4.16) are 0 rather than μ. This is because account must be taken of a bound state in the region between 0 and μ, namely the neutron. It can be produced in π^+ and π^- scattering by means of the diagrams shown in Fig. 4.2. Energy

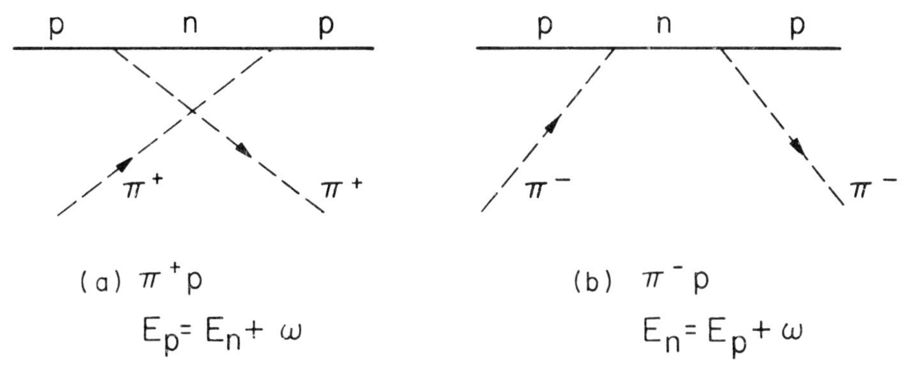

FIG. 4.2 Diagrams responsible for poles in the scattering amplitude. E_p, E_n, and ω are the total energies of the proton, neutron and π meson respectively.

is conserved when,

$$E_p = E_n \pm \omega$$

In the laboratory $E_p = M$, $E_n = \sqrt{k^2 + M^2}$, $\omega = \sqrt{k^2 + \mu^2}$. Substituting and solving for ω gives,

$$\omega = \pm \frac{\mu^2}{2M} \tag{4.17}$$

The + sign is for π^+ scattering and the − sign for π^- scattering. These energies are inacessible, of course, since $|\omega| < \mu$ (the wave number, k, is imaginary).

Account will be taken of these bound states by adding Breit-Wigner resonances to the scattering amplitudes and letting the width approach 0 as a limit. By the uncertainty principle a resonance with 0 width must have infinite lifetime as a bound state should. Thus, we write,

$$f_{\pm}^{B}(\omega) = \lim_{\Gamma \to 0} C \frac{(\mp\Gamma)}{\left(\omega \mp \dfrac{\mu^2}{2M}\right) + i\dfrac{\Gamma}{2}} \tag{4.18}$$

where $f_{\pm}^{B}(\omega)$ are the bound state contributions to the $\pi^{\pm}p$ forward scattering amplitudes respectively and C is an unknown constant. The ± signs in the numerator of eq. (4.18) ensure that eq. (4-14) will be satisfied.

The contributions of $f_{\pm}^{B}(\omega)$ to the dispersion relations will now be calculated explicitly in eqs. (4.15) and (4.16). The reason for doing so is that bound states are not included in the optical theorem relating Im $f(\omega)$ to $\sigma_{tot}(\omega)$. Thus,

$$\operatorname{Im} f_{\pm}^{B}(\omega) = \lim_{\Gamma\to 0} C \frac{\left(\mp\frac{\Gamma^2}{2}\right)}{\left(\omega \mp \frac{\mu^2}{2M}\right)^2 + \frac{\Gamma^2}{4}} \tag{4.19}$$

These expressions must be inserted into the integrals on the right hand side of eqs. (4.15) and (4.16), the integrations performed, and then the limit $\Gamma \longrightarrow 0$ be taken. Clearly only $f_{+}^{B}(\omega)$ will contribute to the integrals in the limit $\Gamma \longrightarrow 0$. The amplitude $f_{-}^{B}(\omega)$ contributes implicitly thru previous use of eq. (4.14) to eliminate the integration over $\omega' < 0$. Since,

$$\lim_{\Gamma\to 0} \frac{\frac{\Gamma^2}{2}}{\left(\omega - \frac{\mu^2}{2M}\right)^2 + \frac{\Gamma^2}{4}} = 2\pi\delta\left(\omega - \frac{\mu^2}{2M}\right) \tag{4.20}$$

where $\delta(x)$ is the Dirac delta-function, the integrations are trivial to carry out. The dispersion relations become,

$$\operatorname{Re}\left[f_+(\omega) + f_-(\omega)\right] - \operatorname{Re}\left[f_+(\mu) + f_-(\mu)\right] =$$

$$2\frac{k^2}{\pi} P\int_{\mu}^{\infty} \frac{\omega'}{k'^2} \frac{\operatorname{Im}\left[f_+(\omega') + f_-(\omega')\right]}{\omega'^2 - \omega^2} d\omega' + \frac{2F^2}{\mu^2} \frac{k^2}{\omega^2 - \left(\frac{\mu^2}{2M}\right)^2}\left(\frac{\mu^2}{2M}\right) \tag{4.21}$$

$$\operatorname{Re}\left[f_+(\omega) - f_-(\omega)\right] - \operatorname{Re}\left[f_+(\mu) - f_-(\mu)\right]\frac{\omega}{\mu} =$$

$$\frac{2\omega k^2}{\pi} P\int_{\mu}^{\infty} \frac{1}{k'^2} \frac{\operatorname{Im}\left[f_+(\omega') - f_-(\omega')\right]}{\omega'^2 - \omega^2} + \frac{2F^2}{\mu^2} \frac{\omega k^2}{\omega^2 - \left(\frac{\mu^2}{2M}\right)^2}$$

where the unknown constant,

$$F^2 = 2C \frac{\mu^2}{\frac{\mu^2}{2M} - \mu^2}$$

must be determined by experiment. The scattering amplitudes in eq. (4.21) now contain no bound states and so can be related to the total cross sections using the optical theorem. The lower limits on the integrals are again μ. If eqs. (4.21) are added and subtracted to solve for $\mathrm{Re}\, f_{\pm}(\omega)$ separately and the optical theorem eq. (4.11) used, the result is,

$$\mathrm{Re}\, f_{\pm}(\omega) = \frac{1}{2}\left(1 \pm \frac{\omega}{\mu}\right) \mathrm{Re}\, f_{+}(\mu) + \frac{1}{2}\left(1 \mp \frac{\omega}{\mu}\right) \mathrm{Re}\, f_{-}(\mu) +$$

$$\frac{k^2}{4\pi^2} P\int_{\mu}^{\infty} \frac{d\omega'}{k'} \left\{ \frac{\sigma_{\mathrm{tot}}^{\pm}(\omega')}{\omega' - \omega} + \frac{\sigma_{\mathrm{tot}}^{\mp}(\omega')}{\omega' + \omega} \right\} \pm \frac{2F^2}{\mu^2} \frac{k^2}{\omega \mp \dfrac{\mu^2}{2M}} \qquad (4.22)$$

These are the final dispersion relations for $\pi^{\pm}p$ scattering. Equations (4.22) were originally derived by Goldberger, Miyazawa, and Oehme [Ref. 6]. They obtained the pole terms from field theory and were able to show that the constant F^2 is approximately equal to the unrationalized renormalized P-wave coupling constant used in the pseudo-vector coupling model.

The derivation of the dispersion relations given above is by no means rigorous. Arguments that were little more than plausibility arguments were used at several points. However, the above derivation is both simple and intuitive. The rigorous derivations given in Refs. 5 and 6 are neither.

Examination of eq. (4.22) shows that the first term in the integral will determine the behavior of $\mathrm{Re}\, f_{\pm}(\omega)$ as a function of ω. All terms except this one are monotonic with energy. Consider the behavior of this term as ω passes through a region where there is a peak in $\sigma_{\mathrm{tot}}^{\pm}(\omega)$. For ω below the peak, this integral will be positive since $\omega' - \omega > 0$ in the region where $\sigma_{\mathrm{tot}}^{\pm}(\omega')$ has a maximum. For ω above the peak, the integral will be negative. If the cross section is symmetric about the peak, then this term will be 0 for ω at the peak. Thus, this integral has the character of the derivative of the total cross section. Since the other terms are not rapidly varying, $\mathrm{Re}\, f_{\pm}(\omega)$ will behave somewhat as the derivative of the total cross section.

Spin-Flip Dispersion Relations. Because the spin-flip amplitude vanishes in the forward direction, the dispersion relations (4.22) necessarily involve only the non spin-flip amplitudes. However, dispersion relations can be derived for the slope of the spin-flip amplitude in the forward direction. Such dispersion relations have been used by Davidon and Goldberger [Ref. 7]. Exact dispersion relations involving the relativistically-invariant scattering amplitudes have been summarized by Chew, Goldberger, Low and Nambu [Ref. 8]. One of these dispersion relations depends primarily on the spin-flip amplitude. They are difficult to use because the integrals involve terms which cannot be measured directly. They must instead be calculated from the phase shifts. Hüper has used these dispersion relations to check the consistency of the various sets of phase shifts that have been published [Ref. 9].

4.2 Proof of the Pomeranchuk Theorem

The dispersion relations, eqs. (4.22), can be used to prove eq. (3.6),

$$\sigma^{+}_{\text{tot}}(\infty) = \sigma^{-}_{\text{tot}}(\infty) \tag{3.6}$$

It is assumed that the range of the π-N interaction remains constant in the limit of infinite energy. This implies $\sigma^{+}_{\text{tot}}(\omega)$ and $\sigma^{-}_{\text{tot}}(\omega)$ approach constants as $\omega \longrightarrow \infty$. It will now be shown that these constants must be the same. The proof consists in showing that if $\sigma^{+}_{\text{tot}}(\infty) \neq \sigma^{-}_{\text{tot}}(\infty)$, then eq. (4.22) implies $\text{Re}\, f_{\pm}(\omega) \propto \omega \ln \omega$ as $\omega \longrightarrow \infty$. But if the range of the interaction is constant, then $|f_{\pm}(\omega)| \propto \omega$ as $\omega \longrightarrow \infty$. The contradiction implies eq. (3.6).

Let $\omega_a >> \mu$ be an energy large enough so that for $\omega > \omega_a$, $\sigma^{\pm}_{\text{tot}}(\omega) = \sigma^{\pm}_{\text{tot}}(\infty)$. Then eqs. (4.22) become,

$$\begin{aligned}\text{Re}\, f_{\pm}(\omega) \cong {} & \pm \frac{1}{2}\frac{\omega}{\mu} \text{Re}\, f_{+}(\mu) \mp \frac{1}{2}\frac{\omega}{\mu} \text{Re}\, f_{-}(\mu) \pm \frac{2F^2}{\mu^2}\omega + \\ & \frac{\omega}{4\pi^2}\int_{\mu}^{\omega_a} \frac{d\omega'}{k'}\left\{\sigma^{\pm}_{\text{tot}}(\omega') - \sigma^{\pm}_{\text{tot}}(\omega')\right\} + \\ & \frac{\omega^2}{4\pi^2} P\int_{\omega_a}^{\infty} \frac{d\omega'}{\omega'}\left\{\frac{\sigma^{\pm}_{\text{tot}}(\infty)}{\omega' - \omega} + \frac{\sigma^{\mp}_{\text{tot}}(\infty)}{\omega' + \omega}\right\}, \quad \omega >> \omega_a\end{aligned} \tag{4.23}$$

If the second integral on the right hand side of eq. (4.23) is carried out the result is:

$$\frac{\omega}{4\pi} \ln\left(\frac{\omega}{\omega_a}\right)\left\{\sigma^{\pm}_{\text{tot}}(\infty) - \sigma^{\pm}_{\text{tot}}(\infty)\right\} \tag{4.24}$$

Thus, this term is $\propto \omega \ln \omega$. All the other terms on the right hand side of eq. (4.23) are $\propto \omega$.

From eq. (2.20),

$$f_{\pm}(\omega) = \frac{1}{k}\sum_{\ell=0}^{\infty}\left\{(\ell+1)\, a^{\pm}_{\ell,\ell+1/2}(\omega) + \ell a^{\pm}_{\ell,\ell-1/2}(\omega)\right\} \tag{4.25}$$

The maximum angular momentum $L_{\pm}$ that can contribute to $f_{\pm}(\omega)$ is given by,

$$L_{\pm} = kR_{\pm}$$

where $R_{\pm}$ is the assumed finite range of the $\pi^{\pm}$- p interaction respectively.

Since $|a_{\ell,\ell\pm 1/2}(\omega)| \leqslant 1$, we can write

$$|f_{\pm}(\omega)| \leqslant \frac{1}{k}\sum_{\ell=0}^{L_{\pm}}(2\ell) \cong \frac{L^2_{\pm}}{k} \cong \omega R^2_{\pm} \tag{4.26}$$

where it has been assumed that $L_{\pm} >> 1$ and $\omega >> \mu$. Equation (4.26) implies that $\mathrm{Re}\, f_{\pm}(\omega)$ cannot contain terms $\propto \omega \ln \omega$. Thus, the term (4.24) must vanish. This in turn implies eq. (3.6). Q.E.D.

4.3 Comparison with Experiment

From Figs. 3.4 and 3.5 it can be seen that the resonance at 200 MeV dominates the behavior of the total cross sections from 0 to about 300 MeV. Thus, we expect $\mathrm{Re}\, f_{\pm}(\omega)$ to be positive up to 200 MeV, go through 0 at about 200 MeV, and become negative above 200 MeV. This behavior was borne out by the first calculations made by Anderson, Davidon, and Kruse [Ref. 10]. This calculation places an important restriction on the scattering amplitude and hence the phase shifts. If we know the phase shifts, then from eqs. (2.20) we can calculate $\mathrm{Re}\, f_{\pm}(\omega)$.

$$\mathrm{Re}\, f_{\pm}(\omega) = \frac{1}{2k} \sum_{j=|\ell-1/2|}^{\ell+1/2} \sum_{\ell} (j + \tfrac{1}{2}) \sin \{ 2\delta^{\pm}_{\ell j}(\omega) \} \tag{4.27}$$

The above authors were, in fact, able to use the dispersion relations to select one out of two possible sets of phase shifts.

Expression (4.22) can be directly compared with experiment. For the differential cross section at 0 deg. one has,

$$\frac{d\sigma(0^\circ)}{d\Omega} = [\mathrm{Re}\, f(\omega)]^2 + [\mathrm{Im}\, f(\omega)]^2 \tag{4.28}$$

The second term on the right can be deduced using the optical theorem, eq. (4.11). Then by measuring the differential cross section at 0 deg., one can deduce the magnitude of $\mathrm{Re}\, f(\omega)$.

The first calculation by G. Puppi and A. Stanghellini seemed to indicate serious difficulties in fitting the experimental values of $\mathrm{Re}\, f_{\pm}(\omega)$ to expression (4.22) [Ref. 11]. In particular, the same value of the coupling constant F^2 did not give a good fit to both the $\pi^+ p$ and the $\pi^- p$ forward scattering amplitude. This was primarily due to the difficulty in fitting data at kinetic energies of 150 and 170 MeV [Ref. 12]. The value of F^2 deduced from $\pi^+ p$ results agreed with other determinations, namely, $F^2 \simeq 0.08$. The value deduced by Puppi and Stanghellini from the $\pi^- p$ results, however, was $F^2 = 0.04$. It was pointed out by M. H. Zaidi and E. L. Lomon that the values of $\mathrm{Re}\, f_-(\omega)$ deduced from eq. (4.22) were extremely sensitive to the slope of $\sigma^-_{\mathrm{tot}}(\omega)$ [Ref. 13]. This is because all the terms in eq. (4.22) for $\pi^- p$ tend to cancel, except the principal value integral, and this is very sensitive to the slope of $\sigma^-_{\mathrm{tot}}(\omega)$.

Schnitzer and Salzman then pointed out that near 200 MeV the determination $\mathrm{Re}\, f_-(\omega)$ was very sensitive to the experimental measurements of σ^-_{tot} [Ref. 15]. In the region 150–250 MeV kinetic energy, where $\mathrm{Re}\, f_-(\omega)$ goes through 0, the use of eq. (4.28) involves taking the difference of two large numbers to deduce one that is small. Specifically

$$|\mathrm{Re}\, f_{\pm}(\omega)| = \left\{ \left(\frac{d\sigma^{\pm}(0,k)}{d\Omega} \right) - (\mathrm{Im}\, f_{\pm}(\omega))^2 \right\}^{1/2} \tag{4.29}$$

Near a resonance $\frac{d\sigma^{\pm}(0,k)}{d\Omega}$ and $|\mathrm{Im}\, f_{\pm}(\omega)|$ become nearly equal. Schnitzer and Salzman state that at 220 MeV "an increase of $\mathrm{Im}\, f_{-}(\omega)$ by 4.5 per cent, which would be caused by a 7.2 per cent increase in the charge-exchange cross section, serves to halve $\mathrm{Re}\, f_{-}(\omega)$." (Recall from Sec. 3.3 that near the 200 MeV resonance the charge-exchange cross section is $\frac{2}{9}$ of $\sigma^{+}_{\mathrm{tot}}(\omega)$.)

Later Schnitzer and Salzman compared eq. (4.22) and experiment in a different way, using the method of least squares to determine the best values of $\mathrm{Re}\, f_{\pm}(\omega)$ and F^2 [Ref. 15]. They demanded that the same value of F^2 fit both π^+p and π^-p data and found a best value $F^2 = 0.08 \pm 0.01$. The value of χ^2 was 23 for the combined fit of π^+p and π^-p data. The expected value for a good fit was 28. The π^-p fit was not as good as the π^+p fit, but was satisfactory.

In 1959, Kruse and Arnold measured the forward amplitude in π^-p scattering at 130 and 152 MeV [Ref. 16]. Their results fell significantly below earlier results [Ref. 12]. This resulted in a substantial improvement in the fit of the calculated forward amplitude to the data [Ref. 17]. By the time of Hamilton's review article, which appeared in 1960, the Puppi-Stanghellini discrepancy was completely resolved [Ref. 18]. The agreement between experiment and eq. (4.22) was excellent.

Currently the viewpoint has returned to that of Anderson, Davidon, and Kruse. Equation (4.22) is accepted as being correct. It is then used to deduce values of $\mathrm{Re}\, f_{\pm}(\omega)$. These values then must be predicted by the phase-shifts from eq. (4.27). Thus, eq. (4.22) supplies an important constraint on any phase-shift analysis.

Fig. 4.3 shows the results of a calculation using recent experimental measurements of the total cross section [Ref. 19]. From 5 BeV to infinity the total cross sections were assumed to be given by the following form [Ref. 20]:

$$\sigma^{\pm}_{\mathrm{tot}} = \sigma_{\infty} + \frac{b^{\pm}}{\sqrt{p}} \tag{4.30}$$

where p is the pion momentum in the laboratory in MeV/c. The following values were used for the constants in eq. (4.30):

$$\sigma_{\infty} = 20.9 \text{ mb}$$

$$b^{+} = 14.0 \text{ mb}/(\mathrm{BeV/c})^{1/2}$$

$$b^{-} = 19.0 \text{ mb}/(\mathrm{BeV/c})^{1/2}$$

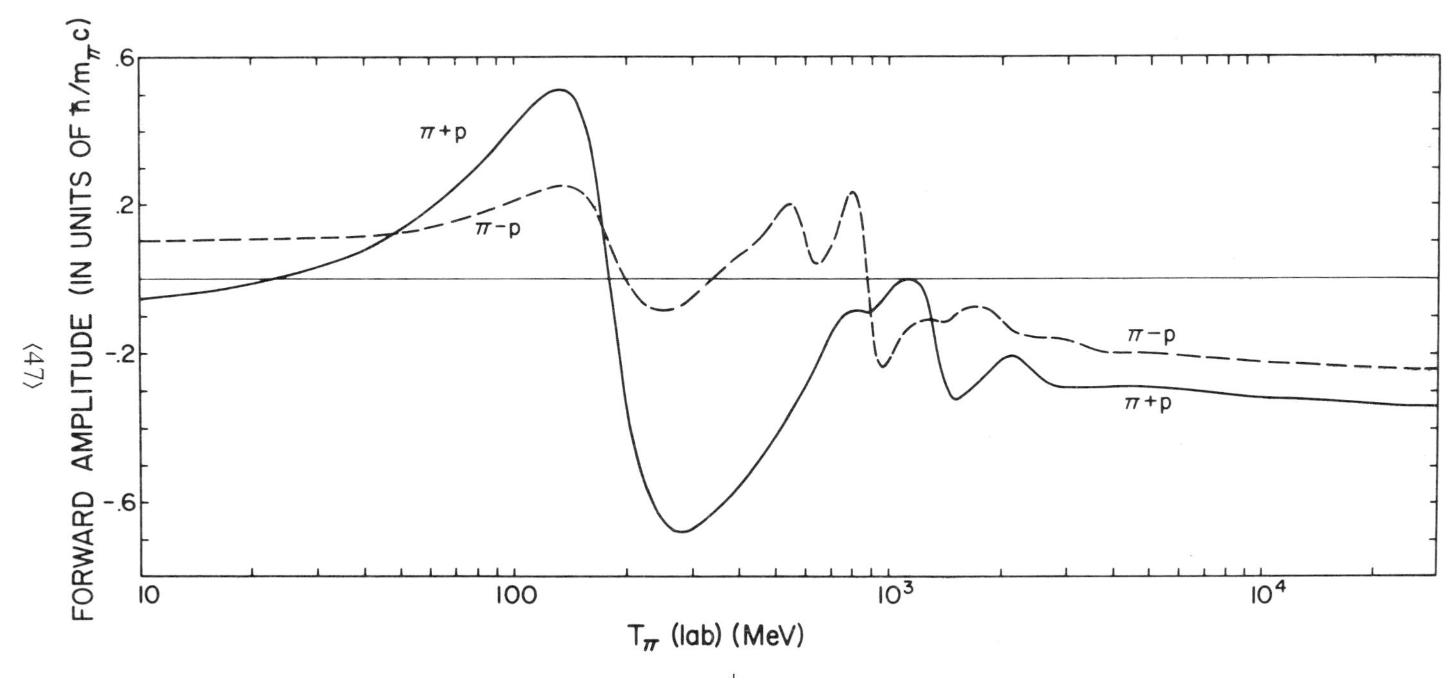

FIG. 4.3 The forward amplitude for $\pi^{\pm}p$ scattering as determined from dispersion theory.

4.4 Forward Amplitude Near a Resonance

In terms of the phase shifts, the forward amplitude can be written

$$kf(\omega) = \sum_{\ell} \sum_{j=\ell\pm 1/2} (j + \tfrac{1}{2}) \frac{\eta_{\ell j}(k) \exp\{2i\delta_{\ell j}(\omega)\} - 1}{2i} \tag{4.31}$$

Höhler has emphasized the usefulness of plotting Re $[kf(\omega)]$ vs. Im $[kf(\omega)]$ [Ref's. 19, 21, 22, 23]. At a well-pronounced resonance, the resonant phase shift is growing rapidly with increasing energy, and as a first approximation, the energy dependence of all other phase shifts can be neglected. Thus, we write

$$kf(\omega) = (j_r + \tfrac{1}{2}) \frac{\eta_{\ell j_r}(\omega) \exp\left\{2i\delta_{\ell j_r}(\omega)\right\} - 1}{2i} + G \tag{4.32}$$

where j_r is the total angular momentum of the resonant state. The first term is rapidly changing due to a resonance. The sum of all non-resonant terms, denoted by G, is assumed to be slowly varying with energy.

For an elastic resonance, $\eta_{\ell j_r} = 1$ and in the plot described above $kf(\omega)$ describes a circle of radius

$$R = \tfrac{1}{2}(j + \tfrac{1}{2}) \tag{4.33}$$

and center $G + iR$. Energy increases along the circle in the counter-clockwise direction beginning at the bottom.[1] Since the resonant phase shift varies more slowly in the lower part of the circle, we expect to see a deformation there caused by the energy dependence of G.

If inelastic processes are present, the energy dependence of η must be considered. This changes the curvature. Fig. 4.4 shows a plot of the resonant term for two cases. The dotted curve corresponds to an elastic resonance and the solid curve to an inelastic resonance where η is monotonically decreasing through the resonance. This kind of resonance has been discussed by Ball and Fraser [Ref. 25].

Fig. 4.5 shows $kf^{(3/2)}(\omega)$ and $kf^{(1/2)}(\omega)$ for πN scattering as calculated by Höhler, Ebel, and Giesecke [Ref. 19]. These are linear combinations of $f_+(\omega)$ and $f_-(\omega)$ with coefficients determined from Table 2.1. Thus,

$$\begin{aligned} kf^{(3/2)}(\omega) &= kf_+(\omega) \\ kf^{(1/2)}(\omega) &= \tfrac{3}{2} kf_-(\omega) - \tfrac{1}{2} kf_+(\omega) \end{aligned} \tag{4.34}$$

where $\pm$ refer as usual to $\pi^{\pm}p \longrightarrow \pi^{\pm}p$ scattering respectively.

$T = \frac{3}{2}$ *Forward Amplitude.* Fig. 4.5 for $kf^{(3/2)}(\omega)$ clearly shows the circular behavior belonging to the first resonance at 200 MeV. The radius is close to 1 unit. From eq. (4.33) we have $j = \frac{3}{2}$ if the resonance is elastic. This agrees

[1] This circular behavior in the pion-nucleon forward scattering amplitude near a resonance was first pointed out by Adair [Ref. 24].

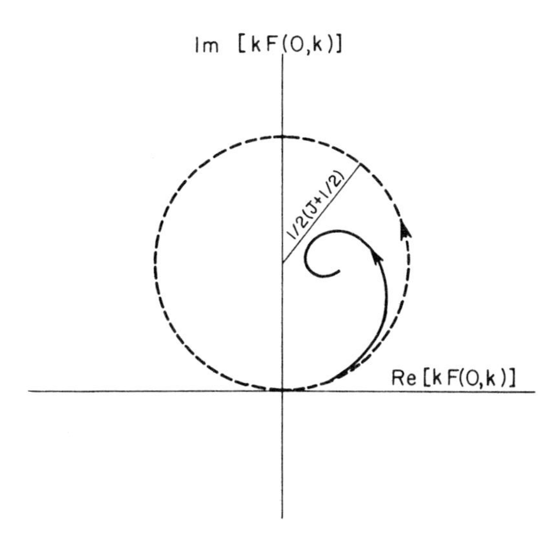

FIG. 4.4 The behavior of the forward amplitude near a resonance. The dotted curve corresponds to an elastic resonance and the solid curve to an inelastic resonance.

with the previous estimate in Sect. 3.3 from the magnitude of the total cross section.

It can be argued that in fact this first resonance must be elastic. The threshold for the first inelastic reaction, single pion production, is almost 175 MeV. Below this energy, then, all $\eta_{\ell j}(k) = 1$. Fig. 4.5 shows that the circle at 200 MeV does not change radius appreciably as one goes from 175 to 300 MeV. Any small change does occur can be attributed to an energy dependent background.

At about 875 MeV a kink is observed in the $T = \frac{3}{2}$ curve of Fig. 4.5. This is undoubtedly a result of the same phenomenon responsible for the shoulder in the $\pi^+ p$ total cross section at about the same energy (see Fig. 3.4).

Near 1200 MeV, we see $kf^{(3/2)}(\omega)$ begin to circle around again. This is due to the resonance at 1300 MeV. In contrast to the 200 MeV resonance, the radius of curvature shrinks as energy increases. This then must be the inelastic type of resonance shown in Fig. 4.4. This means $\eta_{\ell j r}$ is decreasing. This makes it difficult to estimate the appropriate angular momentum. If we set $\eta_{\ell j r} = 1$ this will at least give the minimum possible value for j_r. If the difference between the maximum and minimum of the Re $[kf^{3/2}(\omega)]$ is taken for the diameter of the circle, a value of about 2 units is deduced. From eq. (4.33) $j_r \geqslant \frac{3}{2}$.

$T = \frac{1}{2}$ *Forward Amplitude.* Fig. 4.5 shows two resonances in the $T = \frac{1}{2}$ scattering amplitude. The first one is at 600 MeV. Like the 1300 MeV resonance in the $T = \frac{3}{2}$ state, it is highly inelastic. The radius of curvature of the circle shrinks rapidly above 600 MeV. Using the same criterion as before,

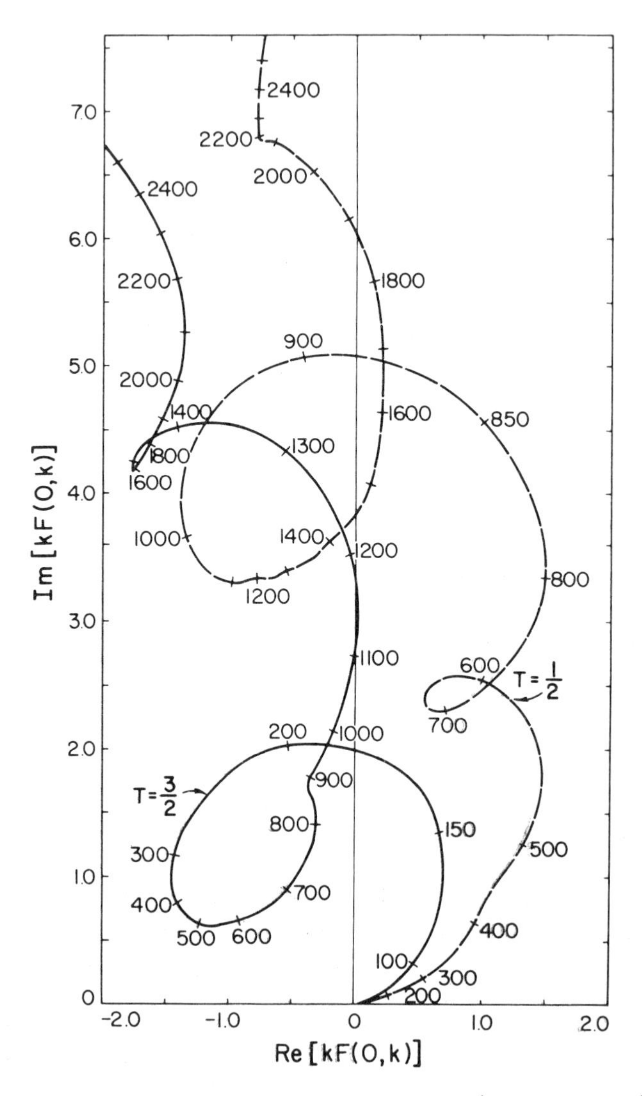

FIG. 4.5 The Im $[kf(\omega)]$ plotted versus Re$[kf(\omega)]$ for $T = \frac{1}{2}$ (- - - -) and $T = \frac{3}{2}$ (——).

we can try to estimate the minimum value of j_r. The radius of curvature then appears to be about $\frac{1}{2}$ which gives $j_r \geqslant \frac{1}{2}$ from eq. (4.33). In this case then, we do not get a useful lower limit.

The second resonance at 900 MeV appears to be almost elastic. The curvature appears to be nearly constant until the energy is well beyond the resonance. The radius of curvature is about $\frac{3}{2}$ units, giving $j_r \geqslant \frac{5}{2}$.

At 2200 MeV there is a sharp kink in the $T = \frac{1}{2}$ forward amplitude curve. This is due to the resonance at that energy. It must be highly inelastic. Also, as can be seen from Fig. 3.5 the resonance is a small effect on a top of a large background. These two observations probably account for the fact that we see only a kink and not a circle in the $T = \frac{1}{2}$ curve of Fig. 4.5.

Above 2500 MeV very little structure in the curves of Fig. 4.5 is expected. This is because resonance effects are very small compared to background terms. Fig. 4.3 bears this out for Re $f(\omega)$ and Figs. 3.4 and 3.5 show that Im $f(\omega)$ will have very little structure also. This is because a great many partial waves contribute to the cross section. A resonance in any one is going to have only a small effect. This is especially true since the very high energy resonances are likely to be highly inelastic.

REFERENCES

1. R. Kronig, *J. Opt. Soc. Am. 12,* 547 (1962). H. A. Kramers, *Atti congr inter fisici* Como *2,* 545 (1927).
2. M. L. Goldberger and K. M. Watson; Chap. 10, *Collision Theory,* John Wiley and Sons (1964).
3. R. Kronig, *Physica 12,* 543 (1946).
4. R. Karplus and M. Ruderman, *Proceedings of the Fifth Annual Rochester Conference,* Interscience Publishers, Inc., (1955) and *Phys. Rev. 98,* 771 (1955).
5. M. L. Goldberger, *Phys. Rev. 99,* 979 (1955).
6. M. L. Goldberger, H. Miyazawa, and R. Oehme, *Phys. Rev. 99,* 986 (1955).
7. W. C. Davidon and M. L. Goldberger, *Phys. Rev. 104,* 1119 (1956).
8. G. F. Chew, M. L. Goldberger, F. E. Low, and Y. Nambu, *Phys. Rev. 106,* 1337 (1957).
9. R. Hüper, *Zeit. für Phys.* 181, 426 (1964).
10. H. L. Anderson, W. C. Davidon, and U. E. Kruse, *Phys. Rev. 100,* 339 (1955).
11. G. Puppi and A. Stanghellini, *Nuo. Cim. 5,* 1305 (1957).
12. J. Ashkin, J. P. Blaser, F. Feiner and M. O. Stern, *Phys. Rev. 101,* 1149 (1956).
13. M. H. Zaidi and E. L. Lomon, *Phys. Rev. 108,* 1352 (L) (1957).
14. H. J. Schnitzer and G. Salzman, *Phys. Rev. 112,* 1802 (1958).
15. H. J. Schnitzer and G. Salzman, *Phys. Rev. 113,* 1153 (1959).
16. U. E. Kruse and R. C. Arnold, *Phys. Rev. 116,* 1008 (1959).
17. H. P. Noyes and D. N. Edwards, *Phys. Rev. 118,* 1409 (1960).
18. J. Hamilton, *Prog. Nucl. Phys. 8,* 143 (1960).
19. G. Höhler, G. Ebel and J. Giesecke, *Zeit. für Phys. 180,* 430 (1964).
20. G. von Dardel, D. Dekkers, R. Mermod, M. Vivargent, G. Weber, and K. Winter, *Phys. Rev. Letters 8,* 173 (1962).
21. G. Höhler and K. Dietz, *Zeit. für Phys. 160,* 453 (1960).
22. G. Höhler and G. Ebel, *Nucl. Phys. 48,* 470 (1963).
23. G. Höhler, *Phys. Letters 10,* 118 (1964).
24. R. K. Adair, *Phys. Rev. 113,* 338 (1959).
25. J. S. Ball and W. R. Fraser, *Phys. Rev. Letters 7,* 204 (1961).

CHAPTER 5

Elastic Scattering and Polarization

5.1 Introduction

In this chapter, we will discuss the differential cross sections and polarization of the recoil nucleon for the three elastic reactions:

$$\begin{aligned} &\pi^+ p \longrightarrow \pi^+ p \quad \text{(i)} \\ &\pi^- p \longrightarrow \pi^- p \quad \text{(ii)} \\ &\pi^- p \longrightarrow \pi^\circ n \quad \text{(iii)} \end{aligned} \tag{1.13}$$

Strictly speaking, reaction (iii) is an inelastic process. However, since the final state contains only two particles, the same formalism as is used to describe elastic scattering can be used. This is why it is loosely referred to as an elastic process.

In Chapter 2, expressions for the differential cross section and polarization were derived for a spin $\frac{1}{2}$ particle scattered by a spin 0 particle. Because the expressions were derived in the barycentric system, there is no distinction between *incident* particle and *target* particle. Both have the same momentum. We reproduce here the final expressions (2.24) and (2.27) derived in Chapter 2. The arguments $\theta, \phi, \theta_i, \phi_i$ and k are suppressed for simplicity.

$$\frac{d\sigma}{d\Omega} = I_0(1 + P_0 \vec{P}_i \cdot \vec{n}_\perp) \tag{5.1}$$

$$\frac{d\sigma}{d\Omega} \vec{P} = 2\mathrm{Re}(f^*g)\vec{n}_\perp - 2\mathrm{Im}(f^*g)\vec{n}_\perp \times \vec{P}_i + (|f^2| - |g|^2)\vec{P} + 2|g|^2(\vec{n}_\perp \cdot \vec{P}_i)\vec{n} \tag{5.2}$$

where,

$$I_0 = |f|^2 + |g|^2, \tag{5.3}$$

$$P_0 = \frac{2\mathrm{Re} f^* g}{|f|^2 + |g^2|}, \tag{5.4}$$

$\vec{P}_i$ is the initial polarization, $\vec{n}_\perp = \dfrac{\vec{k}_{\text{out}} \times \vec{k}_{\text{in}}}{|\vec{k}_{\text{out}} \times \vec{k}_{\text{in}}|}$ is a unit vector perpendicular to the scattering plane, and $\vec{P}$ is the polarization of the scattered nucleon. The quantities f and g are defined by eqs. (2.20).

We consider four cases for $\vec{P}_i$, namely; 1) $P_i = 0$, 2) $\vec{P}_i = \vec{P}_i n_\perp$, 3) $\vec{P}_i = P_i n_\perp \times \vec{n}_z$, and 4) $\vec{P}_i = P_i \vec{n}_z$ where $\vec{n}_z$ is a unit vector along the z-direction. The incoming particles are assumed to be moving in the $+z$-direction. If the (right-handed) coordinate system is oriented so that the scattering plane lies in the y-z plane, then cases 2), 3), and 4) correspond to P_i lying along the $-x$, $+y$, and $+z$ directions respectively.

Case 1). $P_i = 0$. Clearly (5.1) and (5.2) give

$$\frac{d\sigma}{d\Omega} = I_0 \tag{5.5}$$

$$\vec{P} = P_0 \vec{n}_\perp \tag{5.6}$$

Case 2). $\vec{P}_i = P_i \vec{n}_\perp$. Then (5.1) and (5.2) give

$$\frac{d\sigma}{d\Omega} = I_0(1 + P_0 P_i) \tag{5.7}$$

$$\vec{P} = \frac{P_0 + P_i}{1 + P_0 P_i} \vec{n}_\perp \tag{5.8}$$

Case 3). $\vec{P}_i = P_i \vec{n}_z \times \vec{n}_\perp$. We have from (5.1) and (5.2)

$$\frac{d\sigma}{d\Omega} = I_0 \tag{5.9}$$

$$\vec{P} = P_0 \vec{n}_\perp - AP_i \vec{n}_p + RP_i \vec{s} \tag{5.10}$$

Case 4). $\vec{P}_i = P_i \vec{n}_z$. Then,

$$\frac{d\sigma}{d\Omega} = I_0 \tag{5.11}$$

$$\vec{P} = P_0 \vec{n}_\perp + RP_i \vec{n}_p + AP_i \vec{s} \tag{5.12}$$

where $\vec{n}_p$ is a unit vector along the direction of the scattered particle, $\vec{s}$ is a unit vector in the scattering plane and perpendicular to the direction of the scattered particle,

$$\vec{s} = \vec{n}_p \times \vec{n}_\perp \tag{5.13}$$

and A and R are called rotation parameters. They are defined,

$$I_0 A = -(|f|^2 - |g|^2) \sin\theta \pm 2\mathrm{Im} f^* g \cos\theta \tag{5.14}$$

$$I_0 R = (|f|^2 - |g|^2) \cos\theta \pm 2\mathrm{Im} f^* g \sin\theta \tag{5.15}$$

where θ is the scattering angle in the barycentric system. The + and − signs refer to cases (4) and (3) respectively.

The above formulas for $\frac{d\sigma}{d\Omega}$ and $\frac{d\sigma}{d\Omega}\vec{P}$ all depend linearly on 4 quantities, $|f|^2$, $|g|^2$, $\mathrm{Re} f^* g$, and $\mathrm{Im} f^* g$. Thus $|f|$, $|g|$ and the phase difference between f and g can be measured. The absolute phase of f and g cannot be measured because the wave function can always be multiplied by an arbitrary phase factor $e^{i\alpha}$ and all observables will remain unchanged.

The above four quantities can be determined by measuring I_0, P_0, A, and R. *No other independent measurements are possible*. With an unpolarized target or a target polarized along $\vec{n}_\perp$ it is possible to determine only I_0 and P_0. For example, we can measure $\frac{d\sigma}{d\Omega}$ in Case 2) with $\vec{P}_i = \pm P_i \vec{n}_\perp$. That is the differential cross section is measured with the polarization of the target both "up" and "down." Then,

$$\frac{\frac{d\sigma^+}{d\Omega} - \frac{d\sigma^-}{d\Omega}}{\frac{d\sigma^+}{d\Omega} + \frac{d\sigma^-}{d\Omega}} = P_0 P_i \tag{5.16}$$

where the + and - superscripts refer to the signs of the initial polarizations. If $|P_i| = 1$ then the above quantity is the same as the magnitude of the polarization measured with an unpolarized target (Case (1)). Experimentally it is easier to measure P_0 using (5.16) rather than (5.6). This is because with an unpolarized target a second scatter from an analysing target such as carbon is necessary to measure polarization. Using a polarized target one need only measure the differential cross section when $\vec{P}_i = \pm P_i \vec{n}_\perp$. To be useful of course $|P_i|$ must not be too small.

To measure A and R it is necessary to measure the polarization of the scattered particles in the scattering plane when the initial polarization is also in the scattering plane. To measure A we take $\vec{P}_i = \pm P_i \vec{n}_z$ and measure the component of $\vec{P}$ along $\vec{s}$,

$$(\vec{P}^+ - \vec{P}^-) \cdot \vec{s} = A P_i \tag{5.17}$$

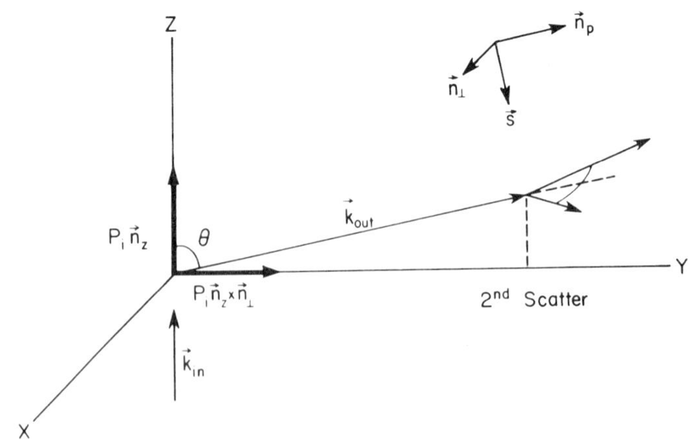

FIG. 5.1 Illustration of the important vectors involved in measuring the A and R rotation parameters.

To measure the polarization along s a second scatter must be performed in a plane perpendicular to the first scattering plane.

In order to measure R we polarize the target so that $\vec{P}_i = \pm P_i \vec{n}_z \times n_\perp$. Then determine

$$(\vec{P}^+ - \vec{P}^-) \cdot \vec{s} = RP_i \tag{5.18}$$

Fig. 5.1 illustrates the important vectors for the measurement of the A and R parameters. We emphasize that the polarizations P^+ and P^- must be determined in the π-N barycentric system. Since they are measured in the lab system of the second scatter, a Lorentz transformation must be performed on the measured polarizations to transform them into the π-N barycentric system.

5.2 Experimental Methods

Measurements have so far been made of I_0 and P_0 only. Most experiments have used unpolarized targets. Recently two experiments have been performed with a target polarized along $\vec{n}_\perp$.

Fig. 5.2 shows a typical set-up for the measurement of the differential cross section using an unpolarized target. A pion beam monitored by three counters

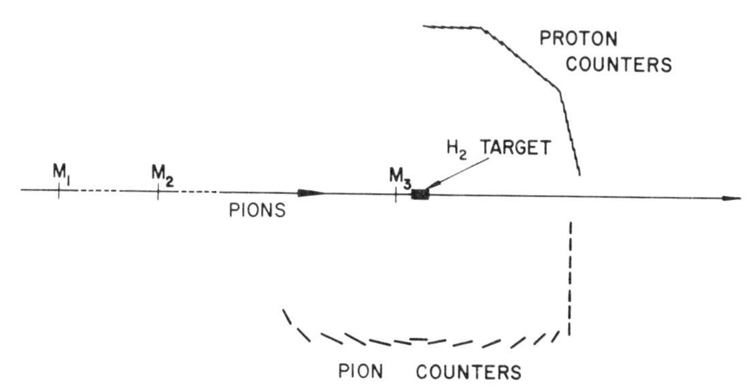

FIG. 5.2 Typical counter set up used for measuring the elastic differential cross section using an unpolarized target.

M_1, M_2, and M_3 is incident on a liquid hydrogen target. The pions are produced by a proton beam from an accelerator and generally have a narrow spread in momentum. A typical value for the full width at half maximum is $\Delta p/p \sim .05$. Proton and pion counters are arranged on opposite sides of the target and in a plane containing the incoming beam line. Coincidences between conjugate pion and proton counters are recorded in coincidence with an incoming pion. By "conjugate" counters we mean a pair of counters positioned so that the two scattered particles from an *elastic* event will pass thru them. For some pairs there is an ambiguity as to which particle is the pion. There are two solutions to this problem. One is to put a Cêrenkov counter in front of the pion counters in order to reject protons. The other is simply to record all the "reverse elastic" pairs and then correct the results accordingly.

The polarization P_0 can also be measured using an unpolarized target. A second scatter must be performed to determine the polarization P_0 produced by the first scatter. We can apply expression (5.7) to the second scatter if the second target has spin 0. Here P_i is the polarization induced by the first scatter and will hence be relabeled P_0. In order to qualify as a Case 1) scatter the plane of the second scatter must be the same as the plane of the first scatter. Then at the second scatterer

$$\frac{d\sigma}{d\Omega} = I_0(1 + \alpha P_0) \tag{5.19}$$

where α is the polarization produced by the second scatterer and P_0 is the polarization produced by the first scatter. Carbon plate spark chambers have been used successfully as the second scatterer.

Recently targets have been developed which contain polarized hydrogen nuclei [Ref. 1]. The polarization is produced by taking advantage of a phenomenon predicted theoretically by Overhauser in 1953 [Ref. 2]. He showed that nuclear orientation could be produced (in metals) by saturating the spin resonance of the conduction electrons. The nuclear polarization results from a spin-spin interaction between electrons and nuclei. It has since been found that this effect occurs in almost any substance containing paramagnetic ions. (For a detailed discussion of the theory see Ref. 3.) The crystal to be polarized is cooled to liquid helium temperature and placed in a strong magnetic field. The electron spin resonance is saturated by irradiating the crystal with an appropriate amount of microwave power. This method is called dynamic nuclear polarization. The crystal used in the π-N scattering experiments is $La_2Mg_3(NO_3)_{12}\cdot 24H_2O$ in which 1 per cent of the Lanthanum has been replaced by even isotopes of Neodymium [Refs. 4, 5]. The hydrogen nuclei in the waters of hydration are polarized. Polarizations as high as 60 per cent have been achieved. Pion interactions are observed using counter hodoscopes. The resolution is made sufficiently good so that π-N interactions can be separated out by taking advantage of two body kinematics. With $\vec{P}_i = \pm P_i \vec{n}_\perp$, measurement of $\dfrac{d\sigma^\pm}{d\Omega}$ determines P_0 thru expression (5.16).

5.3 Parameterization

Ultimately we want to use the differential cross section and polarization data to determine the phase shifts defined in eq. (2.44). This is difficult to do because the experimental quantities are quadratic in the partial wave amplitudes. In addition, these latter are periodic in 2δ. This gives rise to various ambiguities and multiplicities of solutions. These problems will be discussed in some detail in the next chapter.

In order to give some meaning to the experimental results, the data is fit by a linear combination of independent functions. Powers of $\cos\theta$, where θ is the angle of the scattered pion in the barycentric system, have been most often used. However, expansions in a series of Legendre polynomials are gradually becoming more popular. This is because there are certain advantages to using *orthogonal* functions. We will make use of expansions in powers of $\cos\theta$. Thus, we write,

$$\frac{d\sigma(\theta, k)}{d\Omega} = \sum_n a_n(k) \cos^n\theta \tag{5.20}$$

and

$$\frac{d\sigma(\theta, k)}{d\Omega} \cdot P(\theta, k) = \sin\theta \sum_n b_n(k) \cos^n\theta \tag{5.21}$$

The form of (5.21) is suggested by (2.32). The $\sin\theta$ comes from the spin flip term and insures that the polarization will be zero at 0 and 180 deg. These kinds of expansions have the disadvantage that the coefficients a_n and b_n are related to the partial wave amplitudes in a complicated way. This makes their interpretation difficult. However, they do have one clear advantage which makes them useful. The coefficients are easily determined by the method of least squares.

We first discuss the way in which the experimental results are fitted to the expansions (5.20) and (5.21). Following this, the interpretation of the coefficients will be described.

Determination of the Coefficients. The coefficients are determined by the method of least squares. We will briefly describe the method and show how it is used. For a discussion of the theory see, for example, Ref. 6.

Assume a set of n experimental observations O_i with uncorrelated standard deviations Δ_i are taken at angles θ_i where $i = 1, \ldots, n$. Let C be a function of θ and ℓ parameters α_j. Write $C_i = C(\vec{\alpha}, \theta_i)$ where $\vec{\alpha} = (\alpha_1, \ldots, \alpha_\ell)$. We want to find those values of the parameters α_j which give the best fit of the calculated function C to the data. Then probability theory tells us that we must minimize the quantity,

$$\chi^2 = \sum_{i=1}^{n} \left\{\frac{O_i - C_i}{\Delta_i}\right\}^2 \tag{5.22}$$

with respect to the parameters α_j.† The condition for a minimum is evidently,

$$\frac{\partial \chi^2}{\partial \alpha_j} = 0 \qquad j = 1, \ldots, \ell \tag{5.23}$$

This gives ℓ conditions on ℓ parameters. If $C(\vec{\alpha}, \theta)$ is linear in all the α_j, then the eq. (5.23) will be linear in the α_j and there will exist a unique solution. Clearly expansions (5.20) and (5.21) are linear in the coefficients a_n and b_n. This is why they are easily determined.

Thus, the values of the parameters that make χ^2 in (5.22) a minimum can be determined. Probability theory also tells us something about the value of χ^2 at the minimum. Suppose that $C(\vec{\alpha}_I, \theta)$ would be the result of an experiment where $\Delta_i \longrightarrow 0$ for all i. That is, if an exact experiment could be performed with no errors, the curve $C(\vec{\alpha}_I, \theta)$ would be obtained. In a scattering experiment this would mean in general an infinite number of events would have to be recorded. Now define the number of degrees of freedom d,

$$d = n - \ell \tag{5.24}$$

where n equals the number of data points and ℓ equals the number of *independent* parameters α_j. Probability theory tells us that if we calculate χ^2 for a large number of experiments using our ideal function $C(\vec{\alpha}_I, \theta)$, then the mean value of χ^2 will be just equal to d, the number of degrees of freedom [Ref. 8]. Furthermore, the probabilities of obtaining different values of χ^2 for a given value of d can be calculated and are tabulated in tables. If the probability for the minimum χ^2 resulting from a given $C(\vec{\alpha}, \theta)$ is very small, then the function $C(\vec{\alpha}, \theta)$ is an unrealistic one.

The set of all powers of cos θ is a complete set of functions. By taking enough powers of cos θ in (5.20) and (5.21), the results of any experiment can be fitted. We can define a goodness of fit parameter $\sqrt{\chi^2/d}$ which should be about one for a good fit. To determine the number of terms necessary in

†W. C. Davidon has suggested a suitable generalization for the case where there are correlated errors as follows[7],

$$\chi^2 = \sum_{i=1}^{n} (O_i - C_i)\,(H^{-1})_{ij}\,(O_j - C_j)$$

where

$$H_{ij} = \delta_{ij}\Delta_i\Delta_j + \epsilon_i\epsilon_j$$

Here ϵ_i is the correlated error associated with observation 0_i and Δ_i is the uncorrelated error. An example of a correlated error is the normalization error due to uncertainty in beam flux. This error will affect every point in an angular distribution in the same way. It is easy to show that,

$$(H^{-1})_{ij} = \frac{1}{\Delta_i\Delta_j}\left[\delta_{ij} - (\epsilon_i/\Delta_i)(\epsilon_j/\Delta_j)\left\{1 + \sum_{k=1}^{n}\left(\frac{\epsilon_k}{\Delta_k}\right)^2\right\}^{-1}\right]$$

(5.20) and (5.21), the goodness of fit parameter $\sqrt{\chi^2/d}$ is examined for fits involving successively higher and higher powers of cos θ. Enough powers have been included when $\sqrt{\chi^2/d} \simeq 1$. Thus, the number of terms necessary to give a good fit to the data can be determined.

If instead of powers of cos θ a set of orthogonal functions, such as Legendre polynomials, is used in eqs. (5.20) and (5.21), then the coefficients a_n and b_n are approximately independent of the number of terms taken. This is the major advantage of Legendre polynomials over powers of cos θ. The coefficients are not completely independent because eqs. (5.20) and (5.21) are not being fit to an ideal curve but rather to points distributed randomly about some unknown curve. As more terms are added to the expansion, the coefficients of the previous terms may change slightly because they represent projections on a different ideal curve.

Error Matrix. The error matrix gives a measure of the uncertainty in the α_j at their optimum values. It is defined as:

$$M_{ij} = \frac{1}{2} \frac{\partial^2 \chi^2}{\partial \alpha_i \partial \alpha_j} \tag{5.25}$$

evaluated at those values of the α_i and α_j which minimize χ^2.

The uncertainties in the α_j are directly related to M^{-1}, the inverse of M, as follows:

$$\begin{aligned} \overline{\Delta\alpha_i \Delta\alpha_j} &= (M^{-1})_{ij} \\ \Delta\alpha_i^2 &= (M^{-1})_{ii} \end{aligned} \tag{5.26}$$

where the quantities on the left sides of (5.26) are the least-square errors in $\alpha_i\alpha_j$ and α_i, respectively. If M^{-1} has non-zero off-diagonal elements, then pairs of parameters have correlated errors.

Probability theory gives an important constraint on the off-diagonal elements of M^{-1}. Define the correlation coefficient C_{ij},

$$C_{ij} = \frac{(M^{-1})_{ij}}{\sqrt{(M^{-1})_{ii}(M^{-1})_{jj}}} \quad i \neq j \tag{5.27}$$

then we must have,

$$|C_{ij}| \leqslant 1 \qquad \text{all } i \neq j \tag{5.28}$$

If M is diagonal then M^{-1} is diagonal and all $C_{ij} = 0$.

The above ideas concerning the error matrix depend on the assumption that,

$$\frac{\partial \chi^2}{\partial \alpha_i} = 0 \qquad \text{all } i \tag{5.29}$$

That is, χ^2 must be minimized with respect to all the α_i. If an error matrix is determined and it is found that conditions (5.28) are not satisfied, it may be because conditions (5.29) are not satisfied.

Again, if orthogonal functions, such as Legendre polynomials, are used in eqs. (5.20) and (5.21), the off diagonal elements of the error matrix should be considerably smaller. In general, they will not be zero, however.

Interpretation of the Coefficients. In order to interpret the coefficients a_n and b_n in (5.20) and (5.21), it is necessary to relate them to the partial wave amplitudes. This can be done first by substituting explicit expressions for f and g from eqs. (2.20) into eqs. (5.20) and (5.21). Then put in expressions for $P_\ell(\cos\theta)$ and $P_\ell^1(\cos\theta)$ in terms of $\cos\theta$. Perform the indicated squares and collect the coefficients of the various powers of $\cos\theta$. In a similar way the coefficients in an expansion in Legendre polynomials can be determined. The coefficients for expansions in Legendre polynomials and powers of $\cos\theta$ through $\ell = 3, j = \frac{7}{2}$ are displayed in Appendix A.

The expressions are complicated and so the interpretation must be made with some care. The coefficients in (5.20) and (5.21) are most meaningful when they are known as a function of energy.

An important piece of information can be gained by noting the highest even power of $\cos\theta$ in (5.20), the expression for the differential cross section. A study of Table A.3 in Appendix A shows that the highest even power of $\cos\theta$ is just $2j_{\max} - 1$. It is not possible to tell which of the two values of $\ell_{\max} = j_{\max} \pm \frac{1}{2}$ this highest power is due to. If the highest power of $\cos\theta$ is odd, then both values of $\ell_{\max}$ are involved. Note that the odd powers of $\cos\theta$ are always due to interference between states of opposite parity.

It is possible to determine the total angular momentum of a resonance in a similar way. Merely plot the coefficients a_n in (5.20) versus energy in the region of the resonance. Since the elastic cross section goes through a peak, some of the coefficients a_n will show similar behavior. By observing the coefficient of the highest *even* power of $\cos\theta$ exhibiting resonant behavior, the total angular momentum of the partial wave responsible for the resonance can be determined. If n_m is the highest even power, then

$$n_m = 2j_{\text{res}} - 1 \tag{5.30}$$

We have tacitly assumed that a resonance is due to a single partial wave. If this is not true, or if there are two nearby resonances, then the criterion (5.30) may be misleading. It must be used with due caution.

Similar arguments can be made for the polarization. However, polarization experiments are considerably more difficult than scattering experiments. The data are always less accurate. This means that the coefficients b_n in (5.21) cannot be determined as precisely as the coefficients a_n in (5.20). Nevertheless, important conclusions can be drawn from them.

5.4 Experimental Results and Interpretation

In Fig. 5.3 the differential cross sections for $\pi^{\pm}p \longrightarrow \pi^{\pm}p$ and $\pi^{-}p \longrightarrow \pi^{\circ}n$ are displayed at selected energies. Where possible the energies at which resonances occur are included. In Fig. 5.4 the polarization P_0 of the recoil proton in $\pi^{\pm}p \longrightarrow \pi^{\pm}p$ is shown at selected energies.

Figs. 5.5 and 5.6 show the coefficients in eqs. (5.20) and (5.21) as a function of energy for $\pi^{\pm}p \longrightarrow \pi^{\pm}p$. In the next section the energy dependence of these coefficients will be used to learn something about the angular momenta responsible for the various resonances.

At pion energies above 3 GeV elastic scattering appears to be primarily diffraction scattering [Ref. 11]. Between 6 and 20 GeV the differential cross sections for $\pi^{\pm}p \longrightarrow \pi^{\pm}p$ can be fitted to the following simple form [Ref. 12],

$$\frac{d\sigma}{d\Omega} = \frac{\sigma_T(p)}{\sigma_T(p_0)} \exp\left(a + bt + ct^2\right) \tag{5.31}$$

where $\sigma_T(p)$ is the total cross section at pion momentum p, $p_0 = 20$ GeV/c, a, b, and c are constants, and t is the negative square of the four-momentum transfer.

$$t = -2q^2(1 - \cos\theta) \tag{5.32}$$

where q is the momentum of either particle in the barycentric system. The constants a, b, and c have the following approximate values:

$$\begin{array}{lll} \pi^{+}p \longrightarrow \pi^{+}p: & a = 3.60 \pm .05 & \\ & b = 8.8 \pm .3 & (\text{GeV}/c)^{-2} \\ & c = 2.3 \pm .4 & (\text{GeV}/c)^{-4} \end{array}$$

$$\begin{array}{lll} \pi^{-}p \longrightarrow \pi^{-}p: & a = 3.65 \pm .05 & \\ & b = 9.5 \pm .3 & (\text{GeV}/c)^{-2} \\ & c = 2.7 \pm .4 & (\text{GeV}/c)^{-4} \end{array}$$

The 200 MeV Resonance. The 200 MeV resonance will be discussed briefly since it has been treated previously in several textbooks [Ref. 13]. It was argued in Chapter 3 that this resonance was in the isotopic spin state $T = \frac{3}{2}$. Thus, we concentrate on the $\pi^{+}p$ elastic cross section. The differential cross section at 200 MeV has the approximate form $1 + 3\cos^2\theta$. From Table A.3 it is seen that this form results from either pure P_3 or pure D_3 scattering (but not both since there is no $\cos\theta$ term). As mentioned before, Table 1.1 shows that D waves do not become important until almost 400 MeV. So we again conclude that the resonance is in the P_{33} partial wave. Thus, the angular distribution corroborates the deduction made from the total cross section data.

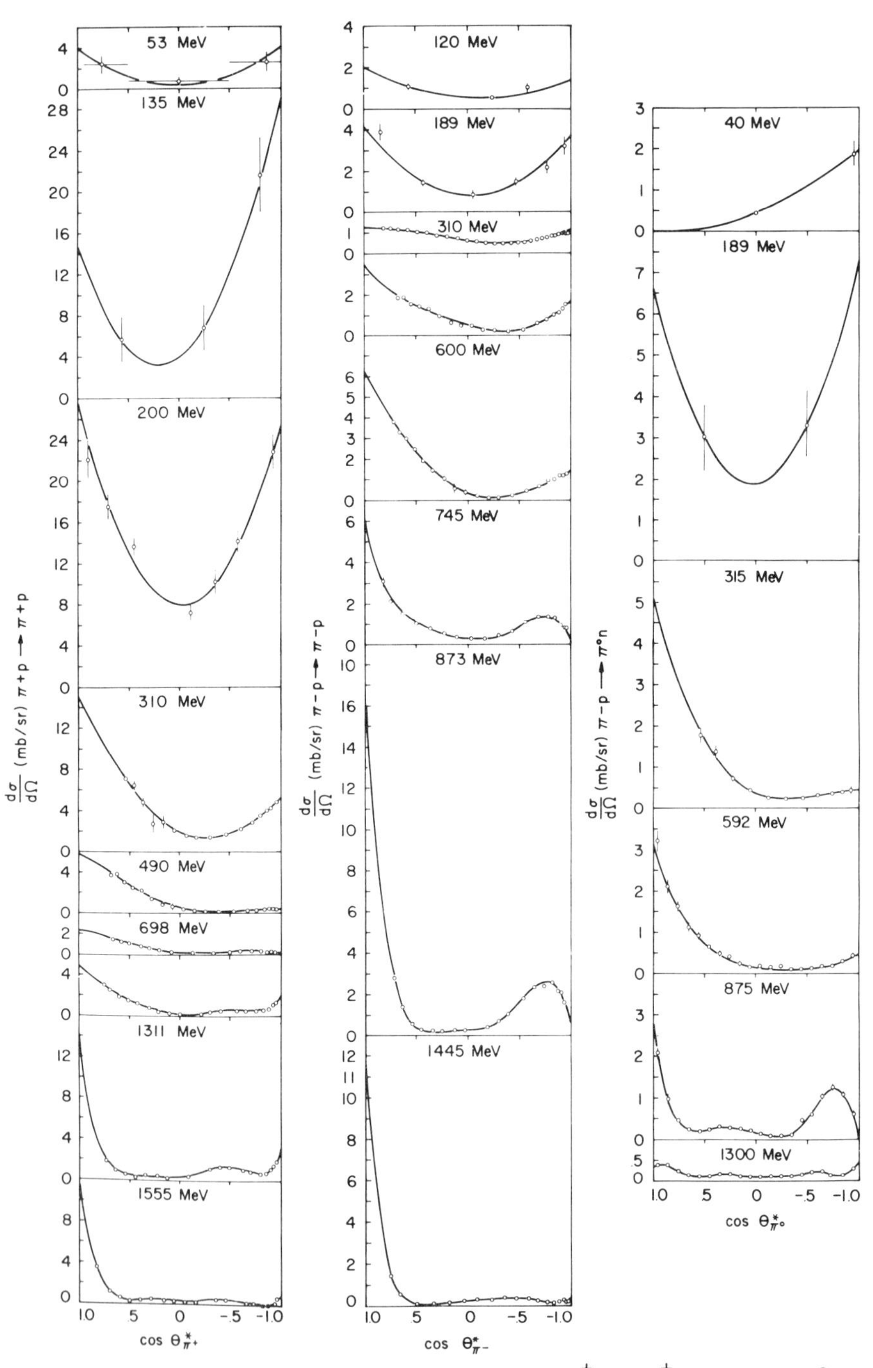

FIG. 5.3 Differential cross sections at selected energies for $\pi^{\pm}p \longrightarrow \pi^{\pm}p$ and $\pi^{-}p \longrightarrow \pi^{\circ}n$. The curves are least square fits to the data. In the higher energy fits a 0 deg. point is calculated from dispersion theory.

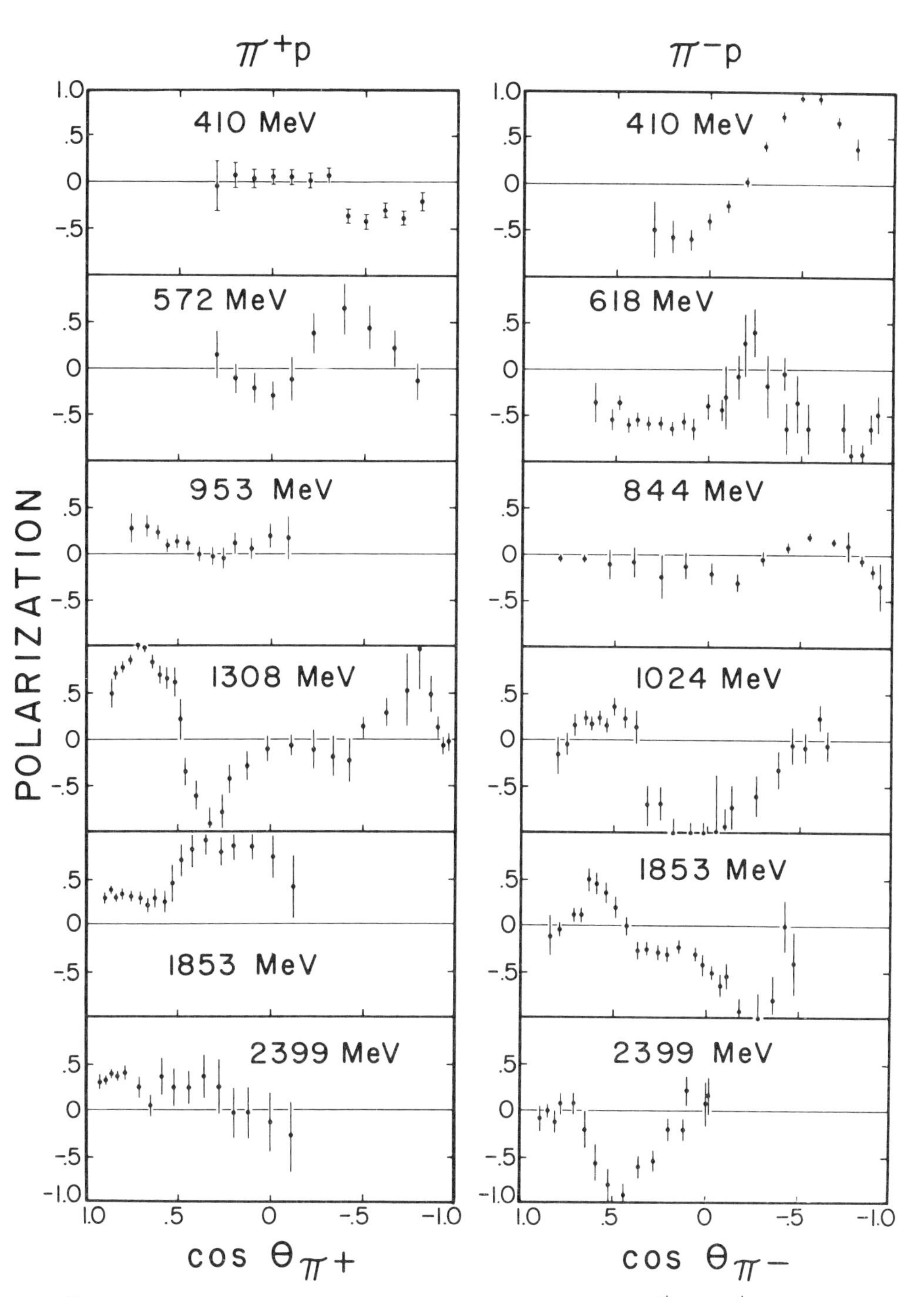

FIG. 5.4 Polarization of the recoil proton at selected energies for $\pi^{\pm}p \longrightarrow \pi^{\pm}p$.

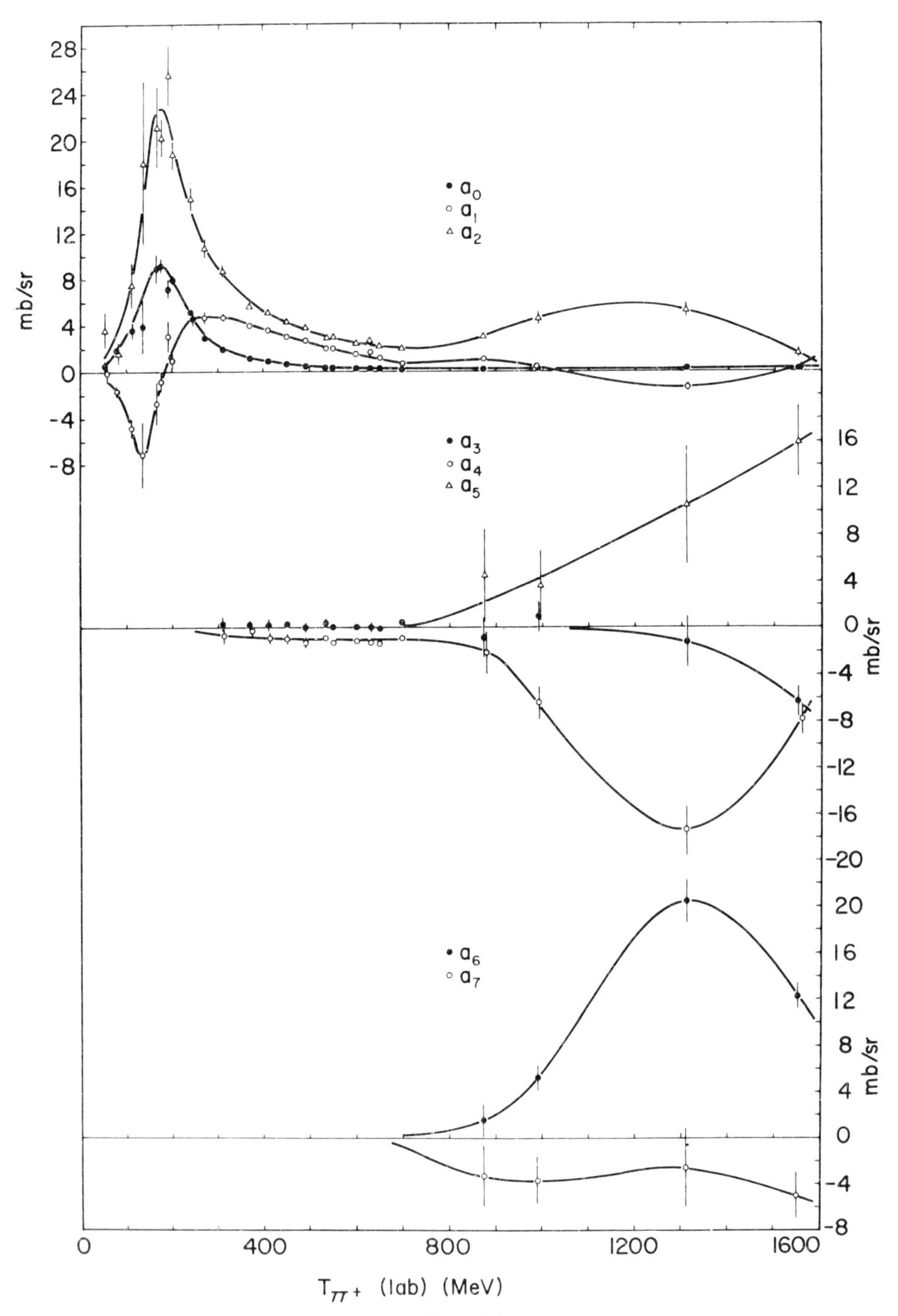

FIG. 5.5 Coefficients in an expansion $\frac{d\sigma}{d\Omega} = \sum a_n \cos^n \theta$ for $\pi^+ p \longrightarrow \pi^+ p$. Curves were fit by eye.

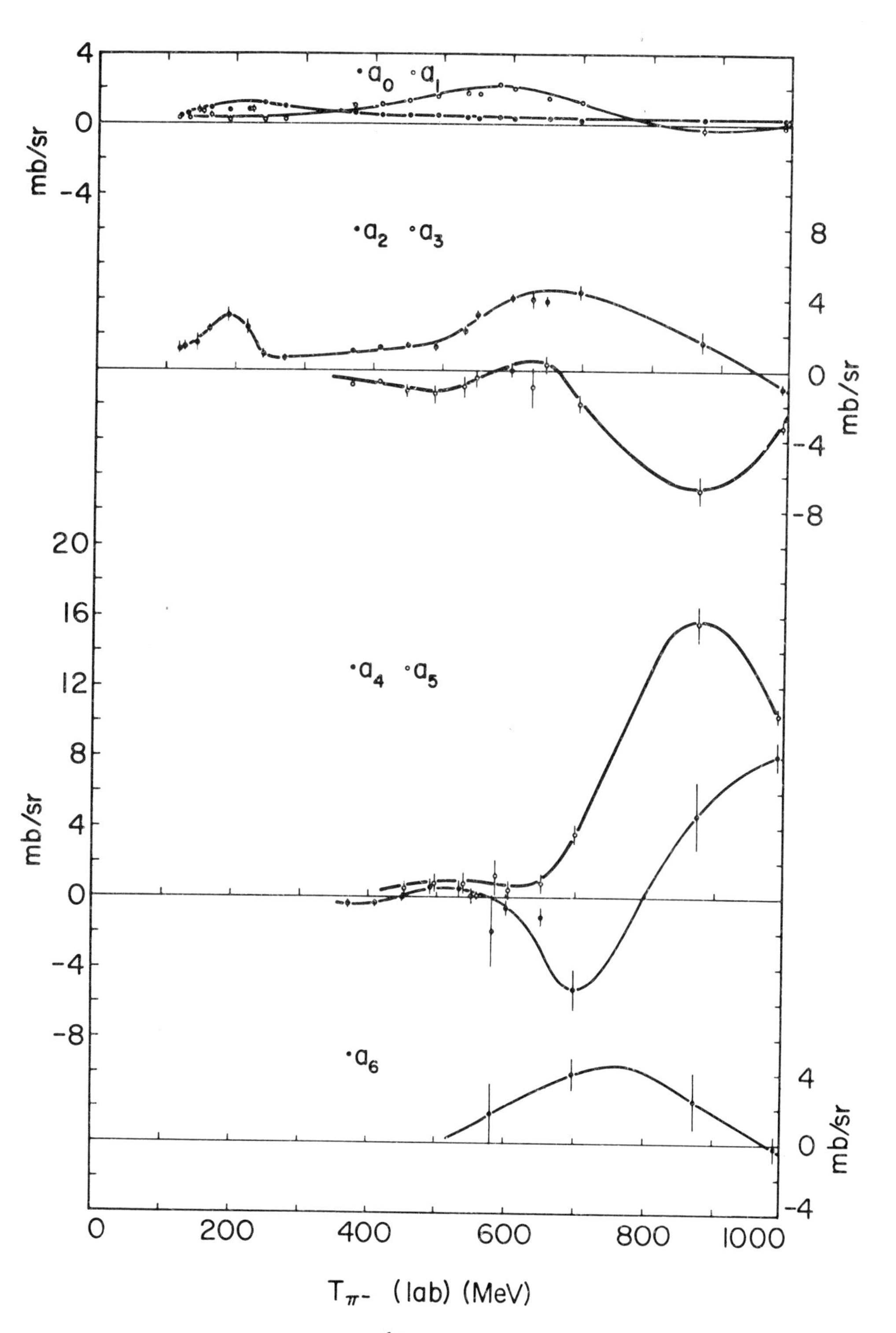

FIG. 5.6 Coefficients in an expansion $\frac{d\sigma}{d\Omega} = \sum a_n \cos^n \theta$ for $\pi^- p \longrightarrow \pi^- p$. Curves were fit by eye.

The 600 MeV Resonance. Fig. 5.6 shows that in the region of the 600 MeV resonance there are peaks in a_0, a_1, a_2 and a suggestion of a peak in a_3 for $\pi^- p \longrightarrow \pi^- p$. Since there is no structure in the coefficients for $\pi^+ p \longrightarrow \pi^+ p$ in this energy region, this must be a $T = \frac{1}{2}$ resonance. The coefficient a_3 does not show a large peak perhaps because it is strongly affected by the tail of the 900 MeV resonance. One would be tempted to say that here we have a $j = \frac{3}{2}$ resonance. The coefficient a_2 is the highest even coefficient showing a peak and a_3, which must be due to an interference between P_3 and D_3 partial waves, also shows a peak. We must be careful, however, to note that a_0, a_1, and a_2 all peak at different energies. The coefficient a_0 peaks at 450 MeV, a_1 at 600 MeV, and a_2 at 650 MeV. A possible explanation is that we have here a $j = \frac{1}{2}$ resonant state interfering with a $j = \frac{3}{2}$ resonant state at a somewhat higher energy.

The 900 MeV Resonance. Fig. 5.6 shows that in the $T = \frac{1}{2}$ state a_3 and a_5 reach dramatic peaks at 900 MeV. Furthermore, a_6 and higher terms are 0. This means $j_{res} = \frac{5}{2}$. Table A.3 then shows that a_5 must be due to a D_5 - F_5 interference. The coefficient $a_3 \simeq - a_5$. This is also compatible with a dominant D_5 - F_5 interference at 900 MeV. This does not imply that both D_5 and F_5 amplitudes are resonant. If $\eta = 1$ and $\delta = 90$ deg. for both the D_5 and F_5 amplitudes at 900 MeV, then we would have $a_5 \cong 135$ mb/sr. This is much greater than the observed 16 mb/sr exhibited in Fig. 5.6. Thus, only one (or neither) of the two amplitudes may be resonant. A detailed analysis of the *energy dependence* of both the a_n and b_n of eqs. (5.20) and (5.21) allows the conclusion that it is the F_5 amplitude that is large [Ref. 4].† The D_5 amplitude, while not negligible, is considerably smaller.

The 1350 MeV Resonance. The peak in the π-N total cross at 1350 MeV appears only in $\pi^+ p$ which allows us to say that this resonance has $T = \frac{3}{2}$. In Fig. 5.5, a_6 has a large peak at 1350 MeV. Higher terms are negligible. Thus we conclude that this is a $j = \frac{7}{2}$ resonance. Again, detailed analysis of the energy dependence of the coefficients permit the conclusion that it is the F_7 amplitude which is resonant or at least dominant.

Resonances above 1500 MeV. No experimental data yet exists which gives an unambiguous determination of the angular momenta responsible for the peaks at 1970 and 3100 MeV in the $\pi^- p$ total cross section.

For two reasons it will be difficult to determine the angular momentum from the powers of $\cos\theta$ expansion. First, the peaks are very small. They amount to no more than 10 per cent of the total cross section. Second, the angular momenta are going to be large. This means that very high powers of $\cos\theta$, up to perhaps $\cos^{10}\theta$, will have to be determined with some precision.

†Actually the polarization is expanded in a sum of $P_n(\cos\theta)$ rather than $P_n^1(\cos\theta)$ in the analysis of Ref. 4.

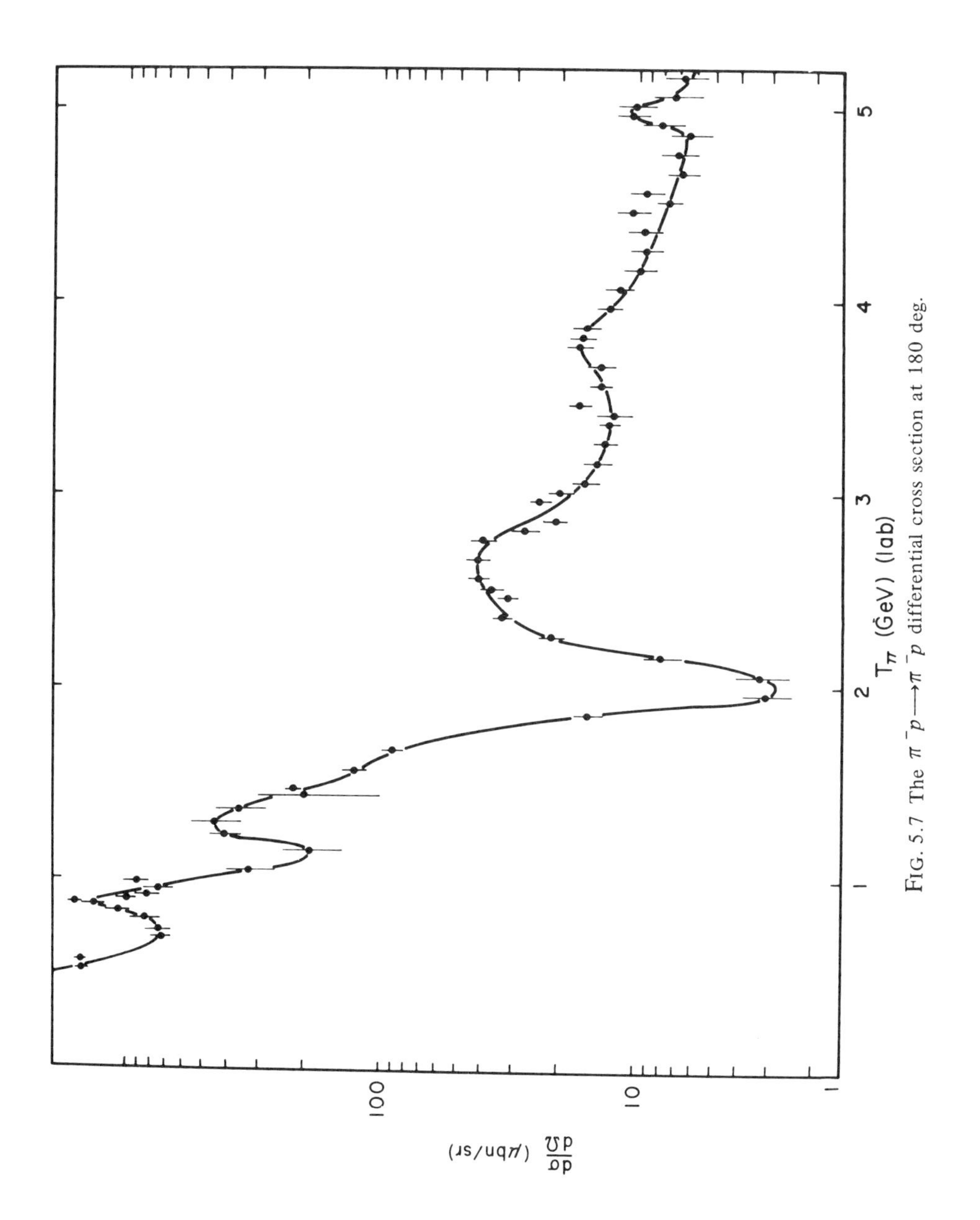

FIG. 5.7 The $\pi^- p \longrightarrow \pi^- p$ differential cross section at 180 deg.

This will be difficult to do. The higher powers of $\cos \theta$ affect the cross section only near 0 and 180 deg.

A possible way to do this would be to measure the charge exchange differential cross sections at many energies near the resonances. As can be seen from Fig. 3.5, the charge exchange cross section is very small above 1 BeV. This indicates that the $T = \frac{1}{2}$ and $T = \frac{3}{2}$ amplitudes nearly cancel each other. Thus a resonance in a single partial wave should make a dramatic change in the differential cross section. To the extent that this is a sensitive test, the cross section overall will be small. So again we have a difficult (but not impossible) experiment.

The $\pi^- p$ differential cross section has been studied at 180 deg. [Ref. 14]. The results are shown in Fig. 5.7 [Ref. 15]. The resonances at 900 and 1350 MeV are observed to interfere constructively with the non-resonant background. On the other hand the resonance at 1970 MeV interferes destructively with the background. If we assume that the amplitude for this background does not change sign as the energy increases from 900 to 1970 MeV, then the 1970 MeV resonance must have parity opposite to the 900 and 1350 MeV resonances. This makes the parity *odd*. Hence, the 1970 MeV resonance must be G_7.

Scattering above 4 GeV. No peaks have been found above 4 GeV in the π-N total cross sections. The angular distributions in the differential cross sections appear to be dominated by the diffraction scattering. Fig. 5.7 shows a peak at $T_\pi = 4.98$ GeV in the differential cross section at 180 deg. This could be due to yet another resonance.

It is quite possible that there are many more resonances above 4 GeV which have so far eluded experiment. This is not surprising. Any resonance above 4 GeV would undoubtedly appear as a very small effect on a very large background. The peaks at 3100 and 3630 MeV in $\sigma_{tot}^{\pm}$ are ~ 0.1 mb above the background. At higher energies the effect can be expected to be even smaller.

REFERENCES

1. O. Chamberlain, C. D. Jeffries, C. H. Schultz, G. Shapiro, and L. van Rossum, *Phys. Letters 7*, 293 (1963).
2. A. Overhauser, *Phys. Rev. 92*, 411 (1953).
3. C. D. Jeffries, *Dynamic Nuclear Orientation*, Interscience Publishers, Inc., (1963).
4. P. J. Duke, D. P. Jones, M. A. R. Kemp, P. G. Murphy, J. D. Prentice, J. J. Thresher, and H. H. Atkinson, *Phys. Rev. Letters 15*, 534 (1965).
5. O. Chamberlain, M. J. Hansroul, C. H. Johnson, P. D. Grannis, L. E. Holloway, L. Valentin, P. R. Robrish, and H. M. Steiner, *Phys. Rev. Letters 17*, 975 (1966).
6. J. Mathews and R. L. Walker, *Mathematical Methods of Physics*, W. A. Benjamin, Inc., (1964), chapter 14. References to other books on probability theory are given at the end of this chapter.

7. M. E. Rose, *Phys. Rev. 91*, 610 (1953).
8. U. E. Kruse and R. C. Arnold, *Phys. Rev. 116*, 1008 (1959).
9. The data displayed in Fig. 5.3 comes from the following experiments: $\pi^{\pm}p \longrightarrow \pi^{\pm}p$. E. C. Fowler, W. B. Fowler, R. P. Shutt, A. M. Thorndike and W. L. Whittemore, *Phys. Rev. 86*, 1053 (L) (1952). A. I. Mukhin, E. B. Ozerov, and B. Pontecorvo, *Soviet Physics* (JETP) *4*, 237 (1960). H. L. Anderson, E. Fermi, R. Martin and D. E. Nagle, *Phys. Rev. 91*, 155 (1953). H. L. Anderson and M. Glicksman, *Phys. Rev. 100*, 268 (1955). H. L. Anderson, W. C. Davidon, M. Glicksman and U. E. Kruse, *Phys. Rev. 100*, 279 (1955). H. R. Rugge and O. T. Vik, *Phys. Rev. 129*, 2300 (1963). J. A. Helland, C. D. Wood, T. J. Devlin, D. E. Hagge, M. J. Longo, B. J. Moyer, and V. Perez-Mendez, *Phys. Rev. 134*, B1062 and B1079 (1964). P. M. Ogden, D. E. Hagge, J. A. Helland, M. Banner, J. Detoeuf, and J. Teiger, *Phys. Rev. 137*, B1115 (1965). P. J. Duke, D. P. Jones, M. A. R. Kemp, P. G. Murphy, J. D. Prentice, and J. J. Thresher, *Phys. Rev. 149*, 1077 (1967). $\pi^{-}p \longrightarrow \pi^{0}n$. J. Tinlot and A. Roberts, *Phys. Rev. 95*, 137 (1954). D. L. Lind, B. C. Barish, R. J. Kurz, P. M. Ogden, and V. Perez-Mendez, *Phys. Rev. 138*, B1509 (1965). C. B. Chiu, R. D. Eandi, A. C. Helmholz, R. W. Kenney, B. J. Moyer, J. A. Poirier, W. B. Richards, R. J. Cence, V. Z. Peterson, N. K. Sehgal, and V. J. Stenger, *Phys. Rev. 156*, 1415 (1967).
10. The data displayed in Fig. 5.4 comes from the following experiments: R. D. Eandi, T. J. Devlin, R. W. Kenney, P. G. McManigal, and B. J. Moyer, *Phys. Rev. 136*, B536 (1964). P. Bareyre, C. Bricman, M. L. Longo, G. Valladas, G. Villet, G. Bizard, J. Duchon, J. M. Fontaine, J. P. Patry, J. Seguinot, and J. Yonnet, *Phys. Rev. Letters 14*, 198 and 878 (1965). Also Refs. 4 and 5.
11. M. L. Perl, L. W. Jones, and C. C. Ting, *Phys. Rev. 132*, 1252 (1963).
12. K. J. Foley, S. J. Lindenbaum, W. A. Love, S. Ozaki, J. J. Russell, and L. C. L. Yuan, *Phys. Rev. Letters 11*, 425 (1963).
13. H. A. Bethe and F. de Hoffman, *Mesons and Fields*, vol. II, Row, Peterson and Co. (1955). W. S. C. Williams, *An Introduction to Elementary Particles*, Academic Press (1961). G. Källén, *Elementary Particle Physics*, Addison-Wesley (1964).
14. S. W. Kormanyos, A. D. Krisch, J. R. O'Fallon, and K. Ruddick, *Phys. Rev. Letters 16*, 709 (1966).
15. The data above 1500 MeV comes from Ref. 14. The data below 1500 MeV comes from Ref. 4 and the seventh experiment listed in Ref. 9. These latter were taken at 165 and 160 deg. respectively.

CHAPTER 6

Phase Shift Analysis

6.1 Introduction

As mentioned in Chapter 2, for each state of total angular momentum and parity there is a phase shift which gives a measure of the scattering in that state. As we will see in Chapter 9, angular momentum and parity play a central role in the various nuclear models and symmetry schemes that have been proposed. Thus, the phase shifts form an appropriate parameterization.[1]

In this chapter we will begin by describing the methods by which phase shifts are determined. Then some of the difficulties in their determination will be discussed. Finally the results of the various analyses that have been carried out will be described.

6.2 Determination of the Phase Shifts

All methods of determining the phase shifts involve the methods of least squares in some way. This was briefly described in Sect. 5.2. With the assumption of charge independence the phase shifts determine uniquely the following quantities:

$\sigma^+_{\text{tot}}, \sigma^-_{\text{tot}}$	the $\pi^{\pm}p$ total cross sections,
$\text{Re}f^+, \text{Re}f^-$:	the real part of the elastic forward scattering amplitude,
$\sigma^+_{\text{in}}, \sigma^-_{\text{in}}$:	$\pi^{\pm}p$ inelastic cross sections,
$\frac{d\sigma^+}{d\Omega}, \frac{d\sigma^-}{d\Omega}, \frac{d^{(0)}}{d\Omega}$:	$\pi^{\pm}p$ and charge exchange elastic differential cross sections (unpolarized targets),
$\vec{P}^+, \vec{P}^-, \vec{P}^{(0)}$:	polarization of the recoil proton in $\pi^{\pm}p$ elastic scattering and of the recoil neutron in charge exchange (unpolarized targets) and
$A^+, A^-, A^{(0)}$:	rotation parameters in $\pi^{\pm}p$ and charge
$R^+, R^-, R^{(0)}$:	exchange scattering

All of the above quantities can be measured experimentally except $\text{Re}f^{\pm}$ which can be determined from dispersion theory. *A* χ^2 (eq. 5.22) can thus be defined between all the quantities that are known experimentally (and $\text{Re}f^{\pm}$) and the predictions from the phase shifts. The problem then is to find the set of phase shifts which makes χ^2 a minimum.

Unfortunately we have no way of making a unique determination of the phase shifts. As pointed out in Chapter 5, if the calculated quantities are

[1] By phase shift we mean both the real and imaginary parts or alternatively the real part and the absorption parameter as defined in (2.44).

linear in the parameters, then they can be determined uniquely. The expressions for the experimental quantities noted above are not linear in the phase shifts. This has two effects. First, it makes the job of determining a best set of phase shifts much more difficult, and second, it gives rise to various ambiguities.

Shortly after physicists became interested in phase shift analysis, high-speed computers were developed. Today virtually all phase shift analysis is performed by elaborate programs using these computers. To illustrate the technique we will describe how a computer program would minimize χ^2 using the "grid method." The program first calculates χ^2 for some arbitrary initial set of phase shifts. Then a single phase shift is incremented by Δ. Again χ^2 is calculated. If the new value of χ^2 is less than the old value, the same phase shift is incremented by the same amount Δ. If not, the phase shift is returned to its original value, incremented by $-\Delta$ and χ^2 is re-calculated. The program continues to increment the phase shift until finally an increment by Δ in either direction increases χ^2. This phase shift is then said to be "optimized." This process is repeated for each of the other phase shifts in turn. A change in one phase shift will change the optimum values for the others. Thus, after the last phase shift has been optimized, the program must go back and re-optimize each of the other phase shifts. This cycle is repeated until it is impossible to increment any phase shift by $\pm\Delta$ without increasing χ^2.

This is a laborious process and takes considerable time even on the best high-speed computers available. Some care must be taken to make the program as efficient as possible. A standard feature of these programs is to begin with a large Δ. After all the parameters have been optimized with respect to the increment Δ, the program decreases Δ to say $\Delta/2$. The phase shifts are then optimized using this new increment. The cycle is repeated for smaller and smaller increments until it is found that the phase shifts are changing by an amount small compared to the accuracy desired. In this case, χ^2 will be found to be decreasing very slowly. The last optimum values of the phase shifts are then the final values.

The "grid method" is the oldest method used to minimize χ^2. Other methods have since been developed in order to try to speed up the process of minimization. One of these is the "gradient method." The gradient of χ^2 with respect to all the phase shifts is calculated. Then χ^2 is minimized with respect to changes along the gradient. A trimming that can be added is to calculate χ^2 at three points along the gradient. Then fit these three points to a function quadratic in the distance along the gradient. This function can then be used to predict the optimum phase shifts.

A third technique is the "ravine method" proposed by Gel'fand and Tsetlin [Ref. 1]. This is a tricky method in which χ^2 is first minimized in two nearby n-1 dimensional parallel planes P_0 and P_1 (see Fig. 6.1). Then the phase shifts are moved a fixed amount along the line joining the two minima

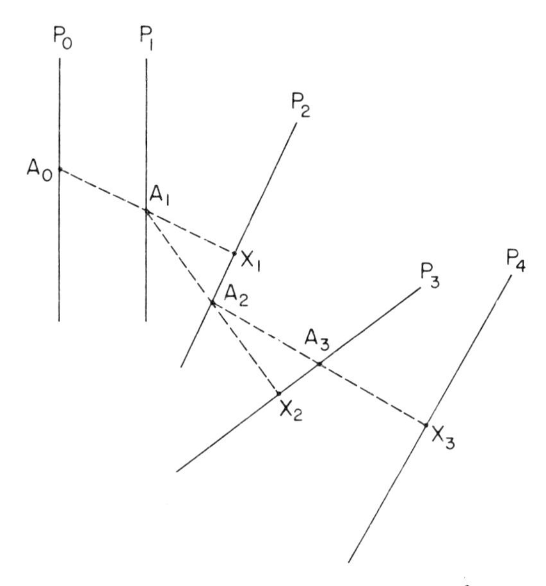

FIG. 6.1 Illustration of the "ravine method" for minimizing χ^2. See text for explanation.

A_0 and A_1. At this point, X_1, a new plane P_2 is defined perpendicular to the line joining the two minima. Now χ^2 is minimized in this plane and a new move by the same fixed amount is made along the line joining this last minimum A_2 and the previous minimum A_1. The effect is to move the phase shifts along "ravines" in χ^2.

The author has personally tried all three techniques including the "quadratic" gradient method. None was found to be appreciably superior to the others. This is undoubtedly due to the fact that the χ^2 surface in an $n + 1$ dimensional hyperspace of n phase shifts and χ^2 is irregularly rippled. This means that, as the program nears the minimum in χ^2, the gradient changes direction rapidly. This could effectively nullify the advantages of the gradient and ravine methods over the grid method.

6.3 Ambiguities

There are several important ambiguities resulting from the non-linearity of the relations connecting the experimental quantities and the phase shifts. Consider, for example, a situation where only the differential cross section is known and the scattering is pure elastic. Also the isotopic spin coordinates will be suppressed for the moment. Suppose terms up to $\cos^{2n}\theta$ are necessary to fit the differential cross section; then $2n + 1$ phase shifts are needed. There are $2n + 1$ *independent* coefficients in the cosine power series which can be determined by measuring the differential cross section at $2n + 1$ points. This is all that can be determined solely by a measurement of the differential cross section. There are $2n + 1$ conditions on $2n + 1$ parameters. If the relations were linear there would be a unique solution. But in fact for the above case Klepikov has shown that there may be as many as 2^{2n+1} valid solutions [Ref. 2].

In order to arrive at a unique solution there must be more conditions than parameters. For the case described above, we might think that measurement of the polarization of the recoil nucleon would be sufficient if precise enough. This is because the different solutions would not give the same polarization. This is almost true. As we shall see later, there would still be two solutions.

We now describe the various ambiguities in detail.

Yang Ambiguity. For the case where there are only S and P waves Yang has shown that if there exists one set of phase shifts, a second set can be found related to the first by:

$$\delta(S_{31}) = \delta'(S_{31})$$

$$\delta(P_{33}) - \delta(P_{31}) = \delta'(P_{31}) - \delta'(P_{33})$$

For a proof of this statement see, for example Ref. 13 of Chapter 5. This ambiguity clearly involves the charge independent description of $\pi - N$ scattering. When D and higher waves become important, this ambiguity disappears. What actually happens is that the two solutions related by the Yang transformation approach each other as D waves come into play.

Minami Ambiguity. This ambiguity, first noticed by Minami [Ref. 3], results from the fact that the differential cross section is invariant under the transformation

$$\delta_{j+1/2,j} \rightleftarrows \delta_{j-1/2,j}$$

From one set of phase shifts, a second set can be generated which will give the same differential cross section if all phase shifts with the same total angular momentum j but with opposite parity are interchanged. Thus,

$$\delta(S_1) \rightleftarrows \delta(P_1)$$

$$\delta(P_3) \rightleftarrows \delta(D_3)$$

$$\delta(D_5) \rightleftarrows \delta(F_5)$$

etc.

The invariance of the cross section is easily seen from the form of the expansion in table A.1 of Appendix A. Likewise from table A.2 it can be seen that the polarization changes sign. Thus, measurement of the polarization resolves this ambiguity.

We note in passing that the forward amplitude, both real and imaginary parts, is invariant under the Minami transformation. This is because in the expression for $f(\theta, k)$ in (2.19), the coefficient of $a_{\ell j}$ depends only on j and not on ℓ, and $g(\theta, k) = 0$ at $\theta = 0$ deg.

Sign Change Ambiguity. If the signs of all the phase shifts are changed, the differential cross section remains unchanged and the polarization changes sign. From one set of phase shifts we can thus generate three more which will give the same differential cross section. One can be generated by changing the sign of all the phase shifts. Using these two and the Minami transformation, two more sets can be generated making four altogether.

Two of these solutions can be eliminated by measuring the polarization. There will still be two left. This is because, if we simultaneously change the sign of the phase shifts and perform the Minami transformation, both the differential cross section and polarization remain unchanged. If the complex phase shift discussed in Sect. (2.6) is used, the combined Minami and sign change transformations can be written as,

$$\delta_{j+1/2,j} \rightleftarrows - \delta^*_{j-1/2,j} \tag{6.1}$$

Under change of sign of the phase shifts (real again) it can be seen from eqs. (2.20) and (2.44), that Imf remains unchanged while Ref changes sign.

The results of the various transformations are shown in Table 6.1. It is seen that under the transformation (6.1) all experimental quantities remain unchanged except Ref and A. Thus, these quantities should be quite helpful in resolving the ambiguity.

TABLE 6.1

Effect of Various Phase Shift Transformations

	$\delta_{j+1/2,j} \rightleftarrows \delta_{j-1/2,j}$	$\delta_{\ell j} \longrightarrow -\delta_{\ell j}$	$\delta_{j+1/2,j} \rightleftarrows -\delta^*_{j-1/2,j}$
$\dfrac{d\sigma}{d\Omega}$	even	even	even
P	odd	odd	even
Imf	even	even	even
Ref	even	odd	odd
A	odd	even	odd
R	even	even	even

Periodicity Ambiguity. One last ambiguity to be mentioned is due to the fact that the scattering amplitude is periodic in δ with period π. This in practice causes no difficulty. As the energy of the pion $T_\pi \longrightarrow 0$, $\lambda \longrightarrow \infty$. If the total cross section at 0 energy is to remain finite, as it does experimentally, then all the partial wave amplitudes must approach 0. The phase shifts must then approach $n\pi$, $n = 0, 1, 2, \ldots$. We adopt the convention that $n = 0$ so that they approach 0 as $T_\pi \longrightarrow 0$. If then the phase shifts are required to be continuous functions of energy, the periodic ambiguity is resolved.

Continuity. There are two techniques for ensuring that the phase shifts will be continuous functions of energy. In one method the phase shifts are determined independently at each energy. They are first determined at the lowest energy for which data is available. Then these phase shifts are used as initial values for the least square analysis at the next higher energy. Better still the phase shifts can be extrapolated to the next higher energy and *these* values used as input at the higher energy. In this way a set of phase shifts is built up which will be a more or less continuous function of energy. Because invariably data is used from more than one experiment, each with its own systematic errors, small phase shifts tend to have a somewhat erratic behavior.

In the second method the phase shifts are determined at all energies simultaneously. The phase shifts are parameterized as a function of energy and a χ^2 is defined which includes the data at all energies. For example, one might use a parameterization suggested by effective range theory and by the threshold dependence of phase shifts for a short range potential ($\tan \delta_\ell$ proportional to $k^{2\ell+1}$)

$$k^{2\ell+1} \cot \delta_{\ell j} = \sum_{n=0}^{N} C_{\ell jn} k^{2n} \tag{6.2}$$

where k is the momentum in the barycentric system. Then χ^2 is minimized with respect to the coefficients $C_{\ell jn}$. This method certainly ensures that the phase shifts will be continuous functions of energy. However, a pitfall of this method is that the parameterization may contain hidden constraints of a strictly mathematical nature. For example, it is not practical to use eq. (6.2) when there is a resonance. So many terms would have to be included that the minimization of χ^2 would take an unreasonably long time. Thus, it is necessary to add specific resonance terms of the form of eq. (2.61). The specific partial waves into which the resonant terms are placed is based on little more than guess work at the higher resonances.

The first of the above methods is called energy-independent analysis and the second energy-dependent analysis.

Uniqueness. From the discussion at the beginning of this section it is clear that it is necessary that there be more independent pieces of information than there are parameters in order to get a unique solution. Just how many more and how precise they must be is not clear. This can only be determined by trial and error. As long as there is more than one solution, additional experiments must be performed. Generally more information is obtained by doing different kinds of experiments rather than adding accuracy to data already obtained. Because of the various ambiguities discussed, a bare minimum is to have data on the total cross section, differential cross section, polarization, and Ref from dispersion theory.

Consider the general situation where there is inelastic scattering. The phase shifts are complex. This doubles the number of parameters. Let us see if the information mentioned in the previous paragraph is sufficient to "over-determine" the problem. Suppose as before terms up to $\cos^{2n}\theta$ are needed to fit the differential cross section. If the data are consistent, terms up to $\cos^{2n-1}\theta$ should be necessary to fit the polarization measurements. There are $2(2n + 1)$ phase shifts and absorption parameters to be determined. Measurement of the differential cross section gives $2n + 1$ independent pieces of information. Polarization data gives $2n$ independent pieces of information. Measurement of the total cross section and a calculation of Ref from dispersion theory gives two more independent pieces of information. Altogether then there are $2n+1 + 2n+2 = 2(2n+1) + 1$ independent pieces of information. This is only one more than the number of parameters. If n is large, all the data would have to be quite precise in order to obtain a unique set of phase shifts. Unfortunately, this qualitative statement is about all that can be said. Generally polarization data have large statistical errors. Because of this a unique set of phase shifts will probably not be obtained with just the above information. Examples will be given later.

So far nothing has been said about the various charge states possible in $\pi - N$ scattering. If only $\pi^{\pm}p \longrightarrow \pi^{\pm}p$ is considered, the scattering can be described by separate set of phase shifts for each reaction. Alternatively phase shifts can be used which describe scattering in isotopic spin states $T = \frac{1}{2}$ and $T = \frac{3}{2}$. The partial wave amplitudes for the two descriptions are related by eq. (2.79). If charge independence is assumed, the latter phase shifts can be used to describe charge exchange $\pi^- p \longrightarrow \pi^\circ n$. With no additional parameters the differential cross section and polarization for charge exchange scattering can be predicted. This will obviously be an aid to the determination of a unique set of phase shifts. There are now many more independent pieces of data which can be measured with no addition in the number of parameters. Thus, it should be possible to determine the phase shifts uniquely with sufficiently precise data. Unfortunately, the polarization data that has been available so far is of limited accuracy (Chapter 5). For this reason there are several sets of phase shifts that fit all the currently available data, including polarization, in the region of the second and third resonances.

6.4 Phase Shift Results

It will be convenient to break up the results of various phase shift analyses into four energy regions: 0-300 MeV, 300–700 MeV, 700–1,000 MeV, and above 1,000 MeV.

0–300 MeV. This region is dominated by the large resonance at 200 MeV which, as previously argued, must be in the state $T = \frac{3}{2}$, $j = \frac{3}{2}$ and probably $\ell = 1$. This is borne out by the phase shift analyses which show that the D-wave phase shifts are very small.

The early phase shift analyses gave a P_{33} resonance near 200 MeV [Ref. 4, 5]. Subsequent analyses showed that indeed all the differential cross section data from 0–300 MeV could be fitted using only S and P waves, with $\delta(P_{33})$ going through 90 deg. at 195 MeV [Ref. 6].

Actually, two solutions were found with resonances in the P_{33} state only. These are referred to as the Fermi solutions. One called Fermi I is characterized by positive S_{11} phase shift and the other called Fermi II has the S_{11} phase shift positive at low energy and then suddenly changing sign at about 170 MeV. From these solutions others can be generated. The Yang transformation and the Minami transformation will each double the number of solutions, making 8 altogether. This is exactly the number predicted by Klepikov (cf. Sect. 6.3) [Ref. 2].

Measurement of the recoil-proton polarization from $\pi^- - p$ scattering at 225 MeV eliminated the Fermi II and the Yang I (derived from Fermi I) solutions [Ref. 7]. The application of forward angle dispersion relations by Anderson, Davidon, and Kruse appeared to exclude all solutions in which P_{33} was not resonant [Ref. 8]. However, it was pointed out by Feld that the forward angle dispersion relations could not be used to eliminate the Yang solutions because they give the same real part of the forward amplitude [Ref. 9]. It was shown by Davidon and Goldberger [Ref. 10] and by Gilbert and Screaton [Ref. 11] that the spin-flip dispersion relations were incompatible with the Yang solutions. Later, Lindenbaum and Sternheimer were able to show that the solutions generated by the Minami transformation from both the Fermi and Yang solutions were also incompatible with the spin-flip dispersion relations [Ref. 12]. No distinction is made between Fermi I and Fermi II in the above calculations since the spin-flip dispersion relations are not sensitive to the small phase shifts.

The overall sign of the phase shifts can be determined by measuring the differential cross section at angles small enough to observe the interference with the Coulomb force. Orear has done this for π^+p scattering at 113 MeV [Ref. 13]. He found a reduced differential cross section near 0 deg. This implies a destructive interference between the Coulomb force and the nuclear force. Since the Coulomb force is repulsive the nuclear force must be attractive. The latter is dominated by the P_{33} state. This allows us to conclude that $\delta(P_{33})$ is positive and hence resolves the sign ambiguity.

This leaves only one solution. It is the one that has $\delta(S_{11})$ positive and $\delta(P_{33})$ passing through 90 deg. at 195 MeV. From a theoretical point of view this is the most attractive solution. The large peak in the $T = \frac{3}{2}$ cross section can be interpreted as a one-level resonance in the $j = \frac{3}{2}$, $T = \frac{3}{2}$ state as previously argued. In the Yang solutions, three resonances are necessary—one in P_{33} state and two in the P_{31} state. This is difficult to understand theoretically.

The phase shifts related by the Minami transformations have large D-waves. This is not compatible with what we know about the range of the $\pi - N$ force.

In Chapter 9 we will discuss the static model of Chew and Low which predicts a single resonance in the P_{33} state.

Once the behavior of the phase shifts has been determined the next logical step is to parameterize them as a function of energy. An appropriate parameterization can be devised from plots of the results of the various energy-independent analyses. This has been done by several workers, most recently by McKinley [Ref. 14]. (This latter work contains a complete bibliography of the phase shift analyses published prior to 1963.) His final results for $T = \frac{3}{2}$ are:

$$\begin{aligned} \tan\delta(S_{31})/k &= -0.10 - 0.036k^2 + 0.003k^4 \\ \tan\delta(P_{31})/k^3 &= (-.13 + 0.072\omega - 0.012\omega^2)/\omega \\ q^3 \cot\delta(P_{33}) &= 4.108 + 0.798k^2 - 0.8337k^4 \end{aligned} \tag{6.3}$$

where $\omega = \sqrt{k^2 + 1}$.

For $T = \frac{1}{2}$ he finds two forms, depending upon whether or not he includes the data at 98, 150, and 170 MeV. The data at these three energies do not appear to follow the trend of the experiments at other energies. With all the data included, McKinley finds:

$$\begin{aligned} \tan\delta(S_{11})/k &= 0.17 - .04k^2 + 0.01k^4 \\ \tan\delta(P_{11})/k^3 &= -0.015 + 0.005k^2 \\ \tan\delta(P_{13})/k^3 &= -0.0035 \end{aligned} \tag{6.4}$$

Excluding data at 98, 150, and 170 he finds:

$$\begin{aligned} \tan\delta(S_{11})/k &= 0.17 + 0.02k^2 \\ \tan\delta(P_{11})/k^3 &= 0.016 \\ \delta(P_{13}) &= \delta(P_{31}) \end{aligned} \tag{6.5}$$

No errors are quoted in eqs. (6.3), (6.4) and (6.5) because the data scattered so badly that a least square analysis was not attempted.

One of the difficulties that McKinley had to face was that the results of the various phase shift analyses were not consistent. That is, the results obtained by various workers scatter far beyond what could be reasonably attributed to statistical uncertainty. This is indicated above by the fact that McKinley felt it necessary to publish two sets of $T = \frac{1}{2}$ phase shifts depending upon whether or not he included certain data.

A more direct procedure would be to parameterize the phase shifts as a function of energy and fit all the data from 0 to a few hundred MeV simultaneously. This can be done with confidence since the general behavior of the phase shifts is known. The truth of this statement depends heavily on the thorough phase shift analysis performed at 310 MeV [Ref. 15].

Two energy dependent analyses have been published [Ref. 16, 17]. One (Ref. 17) is completely phenomenological while the other (Ref. 16) uses dispersion relations to calculate the *D* and *F* waves. The results of the two analyses show good agreement for the *S* and *P* phase shifts. For the *D* and *F* phase shifts the agreement is poor. These phase shifts, however, are very small—a few degrees at 350 MeV.

Characteristically energy dependent analyses show high χ^2/d. For the

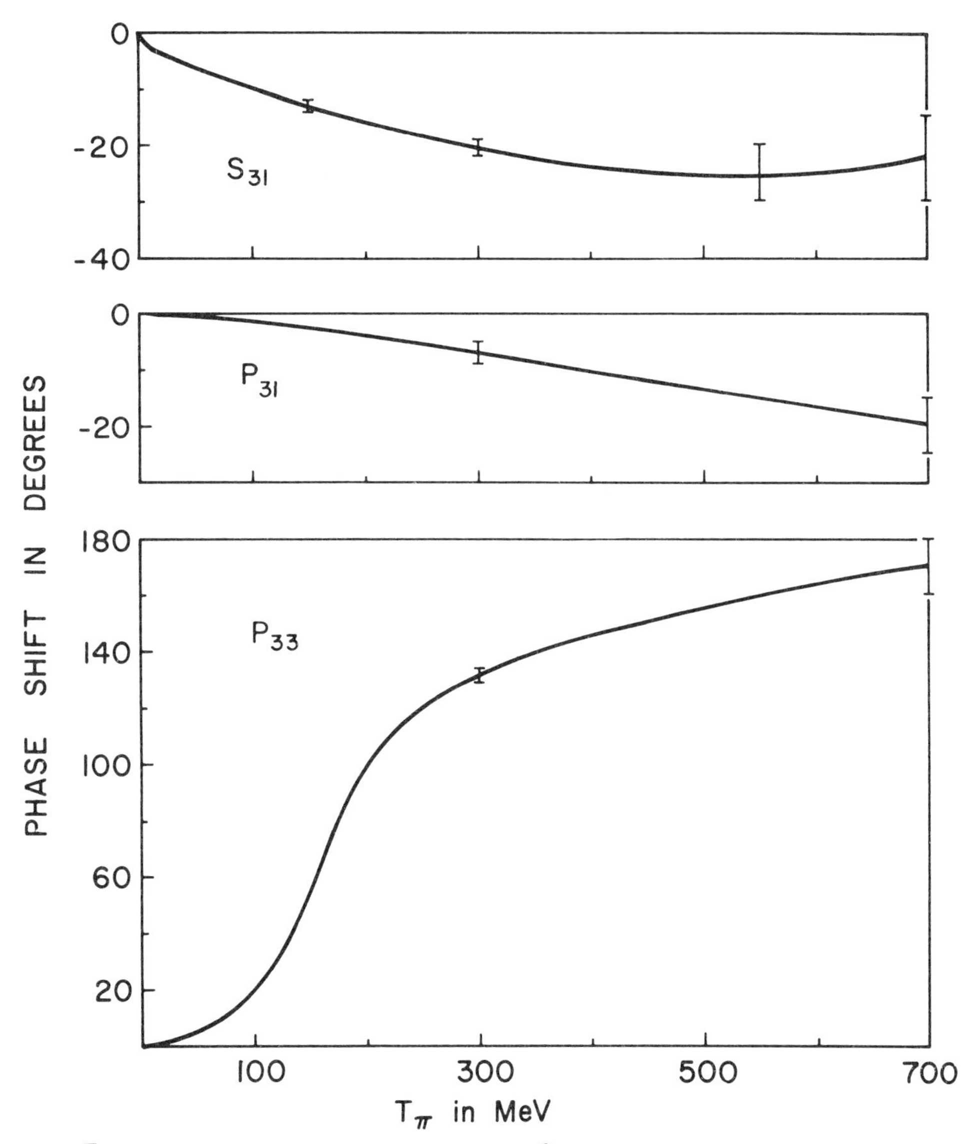

FIG. 6.2 The *S* and *P* wave phase shifts for $T = \frac{3}{2}$. The curves represent an average by eye of the results of all the recent phase shift analyses. The errors shown are estimated from the differences in the various results.

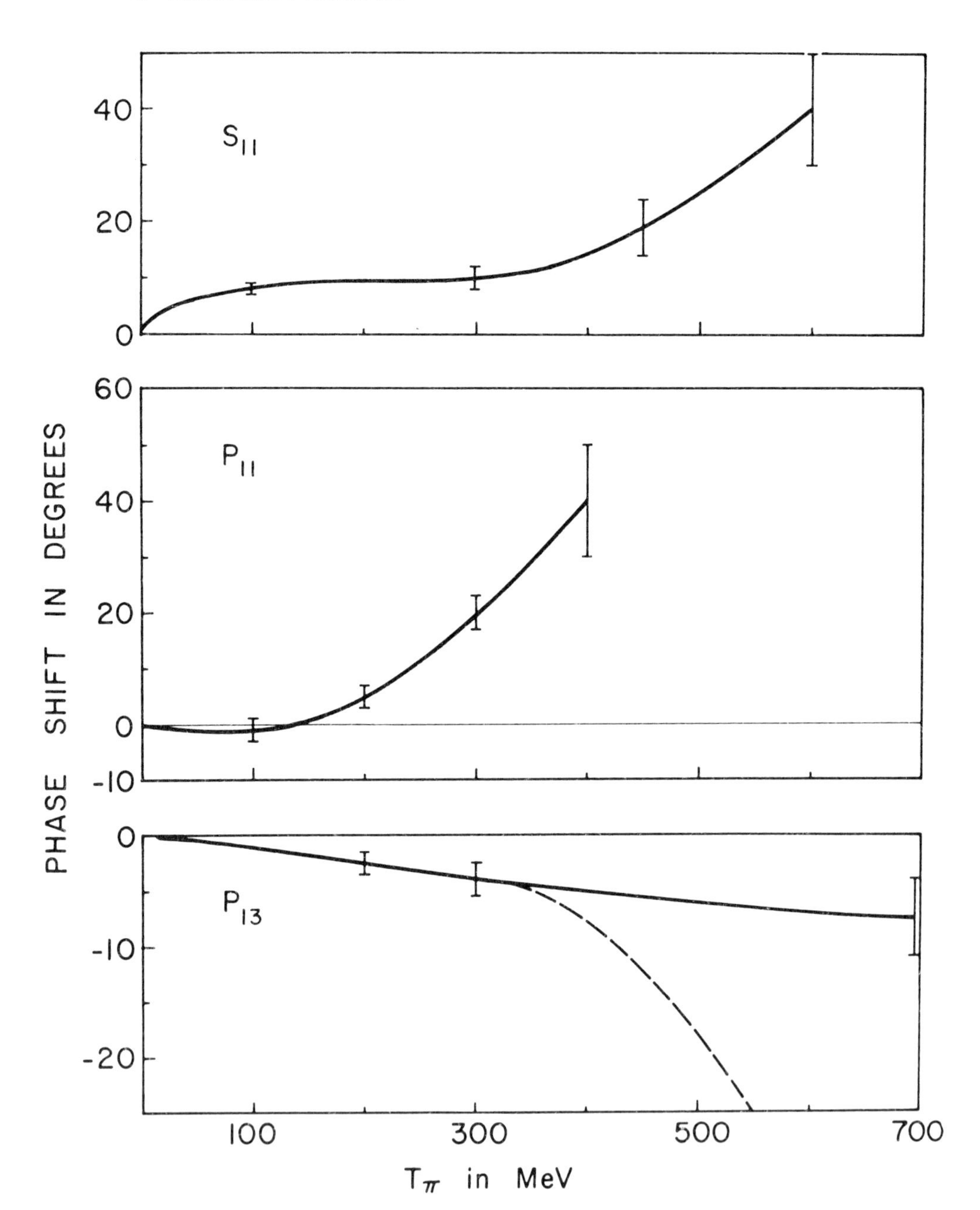

FIG. 6.3 The S and P wave phase shifts for $T = \frac{1}{2}$. The curves represent an average by eye of the results of recent phase shift analyses. The errors shown are estimated from the differences in the various results. The dotted curve for P_{13} is the result of Ref. 21.

above two analyses $d \sim 600$ and $\chi^2/d \cong 2$ for the final results. Statistically the probability is extremely small that χ^2/d would be that large, with so many degrees of freedom. However, data from many different experiments were fit simultaneously. It is probable that the high χ^2/d is due to slight inconsistencies in the results of the different experiments. This is just McKinley's problem in a different form.

Figs. 6.2 and 6.3 show the S and P phase shifts plotted from 0 to 700 MeV. The curves represent a crude average over the results of different phase shift analyses. The error bars are an attempt to estimate the uncertainty due to the variation in results of different analyses. It is in no way related to a standard deviation. The curves up to 350 MeV are based primarily on the results of Refs. 16 and 17.

300–700 MeV. This region is dominated by the resonance in the $T = \frac{1}{2}$ state at 600 MeV. Phase shift analysis above 300 MeV is considerably more difficult because (1) there are more partial waves which need to be included, and (2) there is appreciable inelastic scattering. Because of (2) the phase shifts become complex or alternatively a real phase shift and an absorption parameter must be determined. Either way the number of parameters which must be determined is doubled. For example, if it is necessary to include partial waves through $F_{7/2}$ in both $T = \frac{1}{2}$ and $T = \frac{3}{2}$ states and there is the possibility of inelastic scattering in any partial wave, then there are 28 parameters which must be determined from the data.

In order to assure a unique solution it will be necessary to have more than 28 pieces of independent experimental information. Comparing expressions 5.5 and 5.6 with the tables in Appendix A we find the following number of independent bits of information for each type of experiment:

differential cross sections	π^+p	7
	π^-p	7
	charge exchange	7
polarization	π^+p	6
	π^-p	6
	charge exchange	6
forward amplitude	π^+p	2
	π^-p	2
		43

As of this writing the only data not yet available is polarization of the recoil neutron in charge exchange scattering. Thus, only 37 pieces of information are available. Unfortunately, this doesn't seem to be enough to give a unique solution.

Five groups have published phase shifts in this energy region [Refs. 17-21]. Two of these analyses are energy dependent [Refs. 17, 18]. All sets join the accepted Fermi I solution at lower energies. All are in agreement on the $T = \frac{3}{2}$ S and P phase shifts. There are two reasons for this: (1) because there are no resonances in the $T = \frac{3}{2}$ state in this energy region, the phase shifts are slowly varying and (2) inelastic scattering is very small in the $T = \frac{3}{2}$ state and thus the $T = \frac{3}{2}$ absorption parameters are constrained to remain near 1.0. Except for the fact that they are small (< 10 deg.) there is not much agreement on the D and F phase shifts. Fig. 6.2 shows the $T = \frac{3}{2}$ phase shifts resulting from the various analyses. The meaning of the error bars is the same as above. Table 6.2 summarizes the results of the various analyses for the D and F waves and for the absorption parameters. For $T = \frac{3}{2}$ the table shows that all absorption parameters are near 1. They are not plotted because detailed agreement is poor.

TABLE 6.2

Partial Summary of Phase Shift Results up to 700 MeV

δ	Small Phase Shifts
D_{33}	$<10°$, sign uncertain
D_{35}	$<10°$, negative
F_{35}	$<5°$, sign uncertain
F_{37}	$<5°$, positive
D_{15}	$<10°$, positive
F_{15}	$<10°$, positive and rising
F_{17}	very small, sign uncertain
η	**Absorption Parameters**
S_{31}	$\simeq 1$ up to 600 MeV
P_{31}	$\simeq 1$ (one exception)
P_{33}	$\simeq 1$
D_{33}	$\geqslant .8$
D_{35}	$\geqslant .9$
F_{35}, F_{37}	1.0
S_{11}	falls rapidly above $\eta^{\circ}n$ threshold (560 MeV) to ~ 0.5 at 650 MeV
P_{11}, D_{13}	0.2 - 0.6 at 600 MeV
P_{13}	$>.9$
D_{15}	$\simeq 1$ (one exception)
F_{15}	1.0 (one exception)
F_{17}	1.0

The results of the various analyses for the $T = \frac{1}{2}$ S and P waves are shown in Fig. 6.3. With the exception of the P_{13} phase shift from Ref. 21 the various analyses seem to be in fairly good agreement. Fig. 6.4 shows the D_{13} phase shift. Again the results of Ref. 21 seem to deviate from those of the other analyses. Table 6.2 summarizes the results of the various analyses for the D_{15} and F waves. Also, summarized are the results for the absorption parameters. Let us define, somewhat arbitrarily, a partial wave as being "strongly absorbed" if the absorption parameter in that state is $<.5$. Then two partial

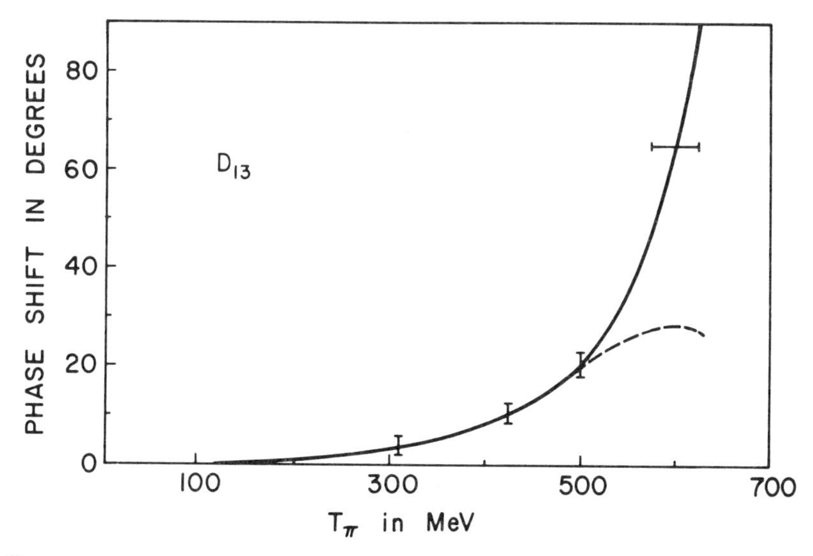

FIG. 6.4 The D_{13} phase shift. The solid curve represents an average by eye of the results of recent phase shift analyses. The errors shown are estimated from the differences in the various results. The dotted curve is the result of Ref. 21.

waves appear to be strongly absorbed, P_{11} and D_{13}. Even the results of Ref. 21 for the D_{13} wave which appears to be an exception have η = .35 at 700 MeV.

In this energy region inelastic scattering is predominantly single meson production. Thus, the two channel approximation discussed in Sect. 2.7 is probably correct. This is especially true when the absorption parameter is close to one up to some energy at which it begins to fall rapidly. It is reasonable that the sudden absorption occurs in one inelastic channel only. This is the situation in both the P_{11} and D_{13} states. The absorption parameter in the P_{11} partial wave is $\approx$1.0 up to ~400 MeV. It then falls to $<$.5 in the next 200 MeV. The rate at which it falls varies considerably in the various solutions. In the D_{13} state the absorption parameter $\approx$1.0 up to 500 MeV and then falls to $<$.5 in the next 100 MeV. The above statements are consistent with *all* the published solutions.

Within the two channel approximation recall that when $\eta <.5$, $\delta_a - \delta_b > 60$ deg. where δ_a and δ_b are the two eigen-phase shifts. At 400 MeV where η is still near 1.0 in the P_{11} partial wave the phase shift is about 40 deg. from Fig. 6.3. Two hundred MeV higher $\eta <.5$, hence the eigen-phase shifts differ by at least 60 deg. If one eigen-phase shift is constant or increases slowly, then the other eigen-phase shift must have gone through 90 degs. slightly below 600 MeV. The Wigner condition excludes the possibility that one eigen-phase shift went rapidly negative. The same argument can be used to show that one D_{13} eigen-phase shift must go through 90 deg. near 600 MeV. These conclusions are consistent with all the published sets of solutions as shown recently by Goldberg [Ref. 22]. We thus have the unusual situation of resonances in both the P_{11} and D_{13} states at about the same energy, a feature first noticed by Roper [Ref. 23]. He found that his (energy-dependent) solution demanded that δ (P_{11}) go through 90 deg. at about 560 MeV with strong absorption. But as observed in Sect. 2.7 when there is strong absorption, the phase shift is irrelevant. A phase shift going through 90 deg. is neither a necessary nor a sufficient condition for a resonance. The argument for a resonance should be based on the strong absorption exhibited. In the solution reported in Ref. 21 no phase shift lay outside ±45 deg. It was argued that, therefore, the data could be fit without any resonances. However, this solution also showed strong absorption in the P_{11} and D_{13} states so that in fact, contrary to the argument, these states may be resonant. The solution presented in Ref. 21 can, however, be discarded because it apparently does not fit the spin-flip dispersion relations [Ref. 24]. Nevertheless, the point remains that widely differing sets of phase shifts still fit the data in this energy region.

700–1,000 MeV. There have been two complete phase shift analyses published in this energy region [Refs. 20, 25] and one which gives only the S_{31} partial wave [Ref. 26]. Two of the analyses (Refs. 20 and 26), but not the third, show sudden strong absorption in the S_{31} partial wave at 800 MeV. The S_{31} phase shift in these analyses falls to ~ - 60 deg. One analysis (Ref. 20) gives sudden absorption in the P_{33} partial wave beginning at about 800 MeV. Recall that there is a shoulder in the $\pi^+ p$ total cross section at about 850 MeV (see Fig. 3.4). Both of the complete analyses give sudden strong absorption in the D_{15} and F_{15} partial waves at 900 MeV. In one analysis (Ref. 25) both D_{15} and F_{15} phase shifts pass through 90 deg. In the other (Ref. 20) the D_{15} phase shift goes no higher than 25 deg. while the F_{15} phase shift goes through 90 deg. The discussion of Section 5.4 favors this latter solution. The strong absorption in these states observed by both these analyses suggest resonances in D_{15} and F_{15} eigen-phases. Two analyses also give an S_{31} resonance at 800 MeV. It is not yet clear that the above results exhaust all the possibilities. Clearly more data will be needed in this energy region in order to resolve the above differences.

Above 1,000 MeV. One group has published a set of phase shifts up to 1,600 MeV [Ref. 20]. It has the following notable features: 1) δ (S_{31}) returns to 0 deg. from ~ - 60 deg. at lower energies, 2) δ (F_{37}) goes through 90 deg. at 1,400 MeV with strong absorption at the same energy ($\eta \approx .25$), 3) δ (S_{11}) passes downward through 90 deg. to almost 0 deg., and 4) δ (D_{13}) and δ (F_{15}) reach 180 deg. Thus, according to this solution, the peak in σ_{tot} at 1,400 MeV is due to an F_{37} resonance.

REFERENCES

1. I. M. Gel'fand and M. L. Tsetlin, *Soviet Physics Doklady 6,* 192 (1961).
2. N. P. Klepikov, *Soviet Physics JETP 14,* 846 (1962).
3. S. Minami, *Prog. Theor. Phys. 11,* 213 (1954).
4. E. Fermi, N. Metropolis, and E. F. Alei, *Phys. Rev. 95,* 1581 (1954).
5. F. De Hoffmann, N. Metropolis, E. F. Alei, and H. A. Bethe, *Phys. Rev. 95,* 1586 (1954).
6. H. L. Anderson and M. Glicksman, *Phys. Rev. 100,* 268 (1955), H. L. Anderson, W. C. Davidon, M. Glicksman and U. E. Kruse, *Phys. Rev. 100,* 279 (1955), J. Orear, *Phys. Rev. 100,* 288 (1955), H. Y. Chiu and E. L. Lomon, *Ann. of Phys. 6,* 50 (1959), V. G. Zinov, S. M. Korenchenko, N. I. Polumordvinova, and G. N. Tentyukova, *Soviet Physics JETP 11,* 1016 (1960).
7. J. F. Kunze, T. A. Romanowski, J. Ashkin, and A. Burger, *Phys. Rev. 117,* 859 (1960).
8. H. L. Anderson, W. C. Davidon, and U. E. Kruse, *Phys. Rev. 100,* 339 (1955).
9. B. T. Feld, *Proceedings of the Sixth Rochester Conference on High-Energy Nuclear Physics,* Interscience Publisher, Inc., (1956), p. 1–19.
10. W. C. Davidon and M. L. Goldberger, *Phys. Rev. 104,* 1119 (1956).
11. W. Gilbert and G. R. Screaton, *Phys. Rev. 104,* 1758 (1956).
12. S. J. Lindenbaum and R. M. Sternheimer, *Phys. Rev. 110,* 1174 (1958).
13. J. Orear, *Phys. Rev. 96,* 1417 (1954).
14. J. M. McKinley, *Rev. Mod. Phys. 35,* 788 (1963).
15. O. T. Vik and H. R. Rugge, *Phys. Rev. 129,* 2311 (1963).
16. M. H. Hull, Jr., and F. C. Lin, *Phys. Rev. 139,* B630 (1965).
17. L. D. Roper, R. M. Wright, B. T. Feld, *Phys. Rev. 138,* B190 (1965).
18. B. H. Bransden, P. J. O'Donnell, and R. G. Moorhouse, *Phys. Letters 11,* 339 (1964) and *Phys. Rev. 139,* B1566 (1965).
19. P. Auvil, C. Lovelace, A. Donnachie, and A. T. Lea, *Phys. Letters 12,* 76 (1964).
20. P. Bareyre, C. Brickman, A. V. Stirling, and G. Villet, *Phys. Letters 18,* 342 (1965) and *Phys. Rev. 165,* 1730 (1968).
21. R. J. Cence, *Phys. Letters 20,* 306 (1966).
22. H. Goldberg, *Phys. Rev. 151,* 1186 (1966).
23. L. D. Roper, *Phys. Rev. Letters 12,* 340 (1964).
24. K. D. Draxler and R. Hüper, *Phys. Letters 20,* 199 (1966).
25. B. H. Bransden, P. J. O'Donnell, and R. G. Moorhouse, *Phys. Letters 19,* 420 (1965).
26. A. Donnachie, A. T. Lea, and C. Lovelace, *Phys. Letters 19,* 146 (1965).

CHAPTER 7

Inelastic Scattering

7.1 Introduction

An enormous amount of data has accumulated on the various inelastic final states initiated by pions on nucleons. In a single chapter we can only cover a small part of these results. We will therefore limit ourselves to experiments most relevant to the pion-nucleon interaction. In particular the emphasis will be on single meson production below 1 GeV. Final states involving strange particles will not be included in the discussion.

Experimentally the inelastic cross sections are dominated by single meson production below 1 GeV. By a meson we mean either π, η, ρ, or ω. The

TABLE 7.1

Meson	Charge States	Spin	Parity	Isospin	G-parity	Mass (MeV)	Width (MeV)
π	$\pm, 0$	0	–	1	–	$139.577 \pm .014$	$\tau^{\pm} = 2.55 \pm .026 \times 10^{-8}$ sec $\tau^{\circ} = 1.78 \pm .26 \times 10^{-16}$ sec
η	0	0	–	0	+	$548.8 \pm .5$	< 10
ρ	$\pm, 0$	1	–	1	+	769 ± 3	112 ± 4
ω	0	1	–	0	–	$782.7 \pm .5$	9.3 ± 1.7

properties of these mesons are summarized in Table 7.1. There are ten such reactions, namely:

$$\pi^+ + p \longrightarrow \pi^+ + \pi^\circ + p \qquad (7.1a)$$

$$\pi^+ + p \longrightarrow \pi^+ + \pi^+ + n \qquad (7.1b)$$

$$\pi^- + p \longrightarrow \pi^- + \pi^\circ + p \qquad (7.1c)$$

$$\pi^- + p \longrightarrow \pi^+ + \pi^- + n \qquad (7.1d)$$

$$\pi^- + p \longrightarrow \pi^\circ + \pi^\circ + n \qquad (7.1e)$$

$$\pi^- + p \longrightarrow \eta + n \qquad (7.1f)$$

$$\pi^- + p \longrightarrow \omega^\circ + n \qquad (7.1g)$$

$$\left\{\begin{array}{l} \pi^+ + p \longrightarrow \rho^+ + p \\ \pi^- + p \longrightarrow \rho^\circ + n \\ \pi^- + p \longrightarrow \rho^- + p \end{array}\right\} \qquad (7.1h)$$

7.2 Experimental Methods

The techniques for measuring the cross sections for reactions 7.1 vary considerably depending on the reaction. Bubble chambers can be used for reactions 7.1a to 7.1d where there are two charged particles in the final state. Counters have also been used. Using counters, an energy distribution of one

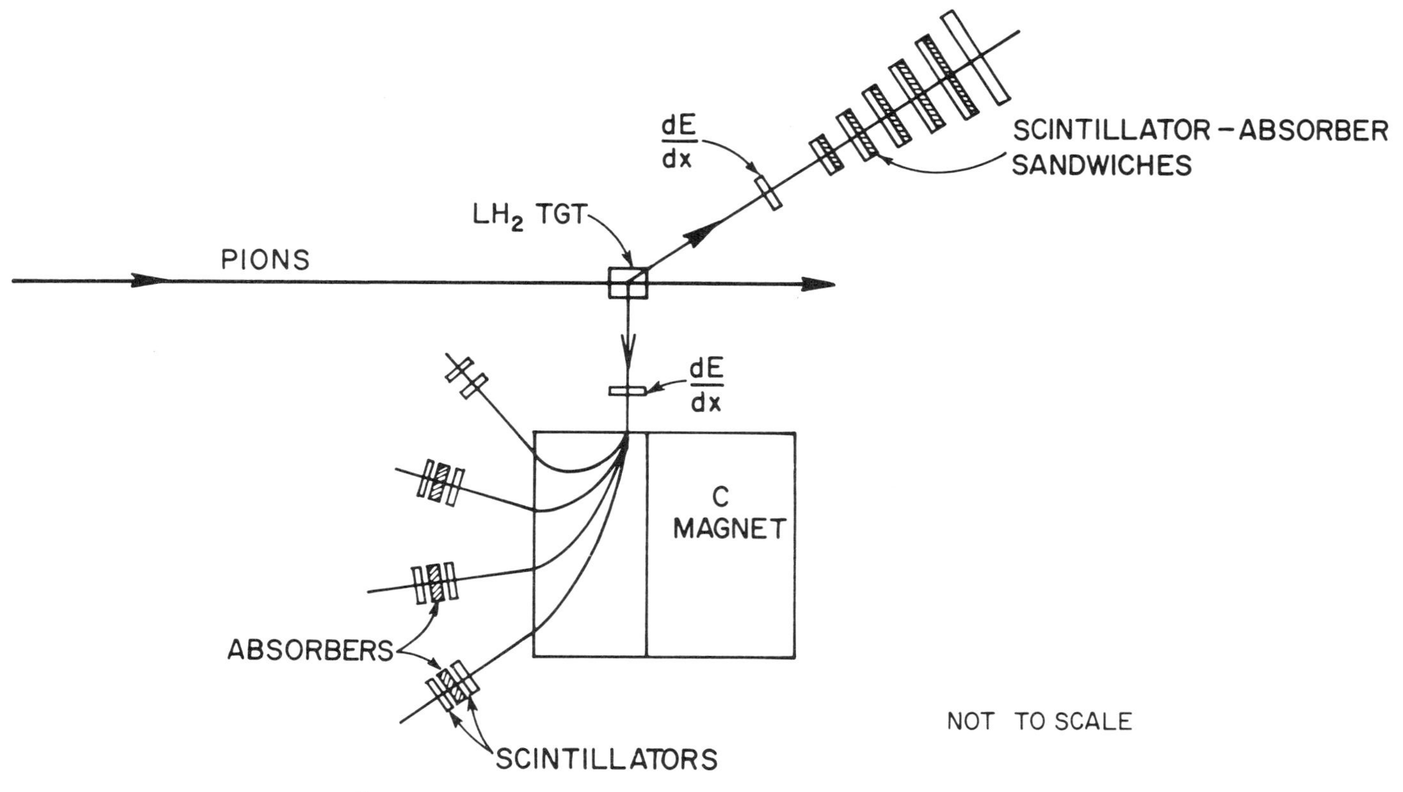

FIG. 7.1 Illustration of two methods of measuring inelastic cross sections.

of the charged particles can be measured at different angles. This will allow inelastic events to be distinguished from elastic events. If positive particles are being detected it is also necessary to measure dE/dX in order to distinguish between protons and positive pions. The energy can be determined by either an arrangement of counters and a magnet or by a series of scintillators and absorbers to measure range [Ref. 1]. At low energies the energy of the final state particle can be determined by measuring its flight time to the counters. The first two methods are displayed in Fig. 7.1. In order to deduce the cross sections for single meson production 7.1a to 7.1e it is necessary to assume that there is no double meson production. For reactions 7.1a and 7.1b the counter sandwich is most appropriate since all charged final-state particles are positive. The cross section for reaction 7.1a can be determined by integrating the energy and angular distribution of final protons. For reaction 7.1b the pion distribution can be measured and then the contribution due to reaction 7.1a must be subtracted. Reactions 7.1c and 7.1d are most conveniently determined by using a magnet to measure the energies of final state proton (7.1c) or π^+ (7.1d). These particles uniquely define the inelastic channel. Reaction 7.1e is difficult to measure. It has been done in three ways: 1) by observing e^+e^- pairs from interactions between an incoming π^- and a bound proton in a propane bubble chamber, 2) by surrounding an LH_2 target with counters to identify events with only neutral final particles and at the same time measuring the time of flight of the neutron [Ref. 1], and 3) by surrounding an LH_2 target with lead-scintillator sandwiches to record the gamma rays from π° decay [Ref. 2]. The last method depends on accurate determination of the efficiency for detection of gamma rays by the counter "telescopes." It also depends on an assumed π° energy distribution.

As can be seen from its width in Table 7.1 the η has a lifetime $\sim 10^{-22}$ sec. Thus, in reaction 7.1f only the decay products will be observed. This reaction has been measured by either completely or partially surrounding an LH_2 target with steel plate spark chambers and observing the decay $\eta \longrightarrow 2\gamma$. This decay can be separated from $\pi^\circ \longrightarrow 2\gamma$ by the much wider angle (in the barycentric system) between the γ's in the η decay.

7.3 Experimental Results

Fig. 7.2 shows the π^-p and π^+p inelastic cross sections up to 1 GeV from a compilation by Rosenberg and Roper [Ref. 3]. The points are calculated by taking the difference between the total and the elastic cross sections. There are several features worth pointing out. (1) The inelastic cross sections are negligible below 300 MeV in spite of the fact that the inelastic threshold for single pion production is at 170 MeV, (2) the π^+p inelastic cross section is much smaller than the π^-p inelastic cross section below 700 MeV, (3) the π^-p inelastic cross section appears to be enhanced at the 600 and 900 MeV

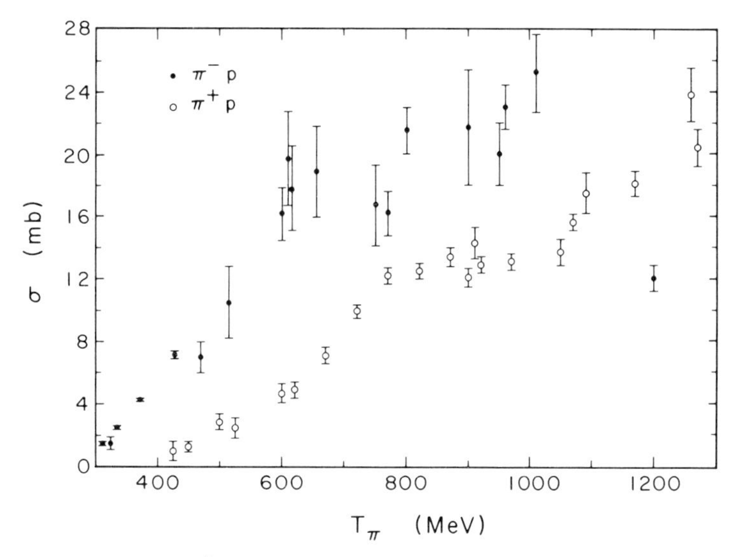

FIG. 7.2 The π^-p and π^+p inelastic cross sections deduced by subtracting the elastic cross section from the total cross section.

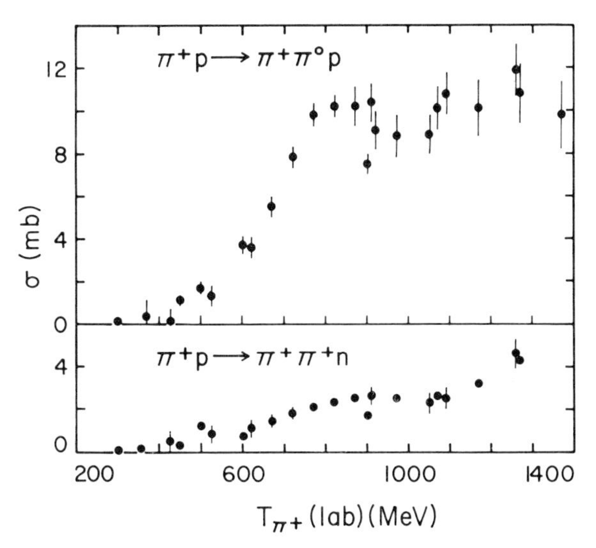

FIG. 7.3 The cross sections for $\pi^+p \longrightarrow \pi^+\pi^0p$ and $\pi^+p \longrightarrow \pi^+\pi^+n$.

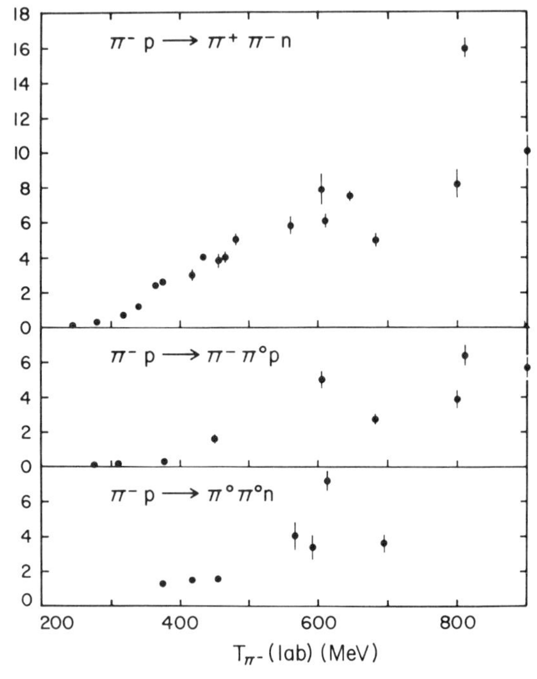

FIG. 7.4 The cross sections for $\pi^- p \longrightarrow \pi^+ \pi^- n$, $\pi^- \pi^\circ p$, and $\pi^\circ \pi^\circ n$.

resonances, and (4) the $\pi^+ p$ inelastic cross section has a shoulder from 800 to 1,000 MeV. In the ensuing discussion we will try to understand some of these features.

Figs. 7.3 and 7.4 show the cross sections for the first five of reactions 7.1 [Ref. 4]. The cross section for 7.1f is shown in Fig. 7.5 [Ref. 5, 6]. There is

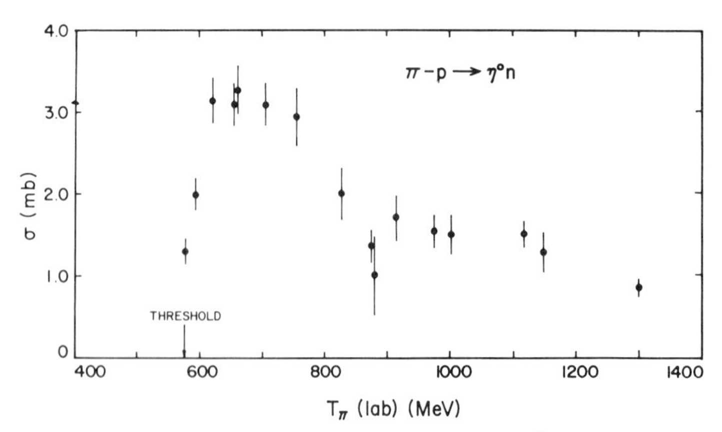

FIG. 7.5 The cross section for $\pi^- p \longrightarrow \eta^\circ n$.

little information on reaction 7.1g below 1 GeV but it is expected to be small since the threshold is at 960 MeV.

What little multiple meson production takes place appears to be primarily due to the decay of the η° into $\pi^+\pi^-\pi^\circ$ or $\pi^\circ\pi^\circ\pi^\circ$. This is evidenced by the sharp rise of the $\pi^- p \longrightarrow \pi^+\pi^-\pi^\circ n$ [Ref. 7] and $\pi^- p \longrightarrow \pi^\circ\pi^\circ\pi^\circ n$ [Ref. 8] cross sections above threshold for reaction (7.1f). Since the ρ meson decays promptly into two pions, reactions (7.1h) can be included in (7.1a) - (7.1e).

7.4 The π-π Interaction. High Energy

At this point we digress slightly to discuss the π-π interaction. A knowledge of this interaction is necessary for an understanding of π-N inelastic scattering.

The π-π elastic cross section is deduced by means of an extrapolation procedure worked out by Chew and Low [Ref. 9]. Consider the process illustrated in Fig. 7.6. We shall refer to this as the one-pion exchange (OPE) process. Let Δ be the four momentum transfer to the proton and ω be the total energy of the two final state pions in their own barycentric frame. The invariants Δ^2 and ω^2 can be written,

$$\Delta^2 = p^2 - [m - (m^2 + p^2)^{1/2}]^2$$

and

$$\omega^2 = [(q^2 + \mu^2)^{1/2} - T]^2 - (q^2 - 2qp \cos\theta + p^2) \tag{7.2}$$

where T and p are, respectively, the kinetic energy and momentum of the recoiling proton in the lab, m the proton mass, μ the pion mass, q the incoming pion momentum in the lab, and θ the lab angle between the incident pion and the recoiling proton.

According to Chew and Low the matrix element for the one-pion exchange process has a pole at $\Delta^2 = -\mu^2$ with a residue given by $FA_{\pi\pi}(\omega^2)$. Here F is the same coupling constant that appears in eq. (4.22) ($F^2 = .08$) and $A_{\pi\pi}$ is the pion-pion scattering amplitude. Precisely we can write,

$$\lim_{\Delta^2 \to -\mu^2} \frac{\partial^2 \sigma_{\mathrm{OPE}}}{\partial(\Delta^2)\partial(\omega^2)} = \frac{F^2}{2\pi}\left(\frac{p}{\mu}\right)^2 \frac{\left[\omega^2\left(\frac{\omega^2}{4} - \mu^2\right)\right]^{1/2}}{(p^2+\mu^2)^2\, q^2}\,\sigma_{\pi\pi}(\omega^2) \tag{7.3}$$

where $\dfrac{\partial^2 \sigma_{\mathrm{OPE}}}{\partial(\Delta^2)\partial(\omega^2)}$ is the differential cross section for the process of Fig. 7.6 and $\sigma_{\pi\pi}$ is the total cross section for the process

$$\pi + \pi \longrightarrow \pi + \pi \tag{7.4}$$

From eq. (7.2) it is clear that in the physical region $\Delta^2 > 0$. Thus, one must *extrapolate* to the point where $\Delta^2 = -\mu^2$. Now the region of small momentum transfer corresponds to small angle, large impact-parameter collisions.

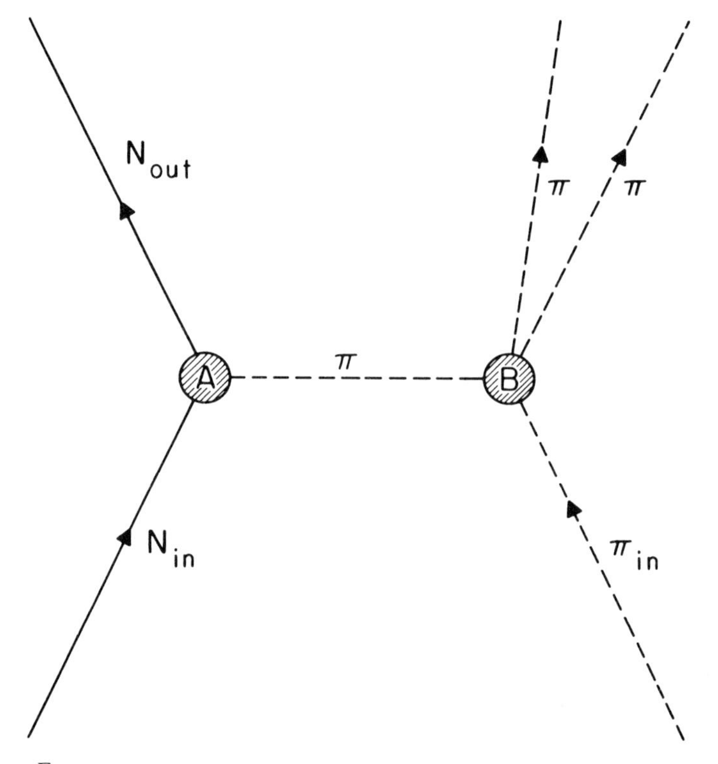

FIG. 7.6 The one-pion exchange diagram for the process $\pi N \longrightarrow \pi\pi N$.

These are the long range "peripheral" collisions. It is reasonable to expect one-pion exchange to dominate in this region.

Unfortunately the situation is not this simple. Because pions have odd parity, single pions are emitted in p-states. This means that the probability of emitting a single pion is zero when it has 0 velocity. This means that

$$\left.\frac{\partial^2 \sigma_{\mathrm{OPE}}}{\partial(\Delta^2)\partial(\omega^2)}\right|_{\Delta^2=0} = 0 \tag{7.5}$$

Thus, at the edge of the physical region the one-pion exchange contribution to $\dfrac{\partial^2 \sigma}{\partial(\Delta^2)\partial(\omega^2)}$ vanishes. This fact makes the extrapolation to $\Delta^2 = -\mu^2$ extremely difficult. Contributions to $\dfrac{\partial^2 \sigma}{\partial(\Delta^2)\partial(\omega^2)}$ due to exchange of more than one pion do not vanish at $\Delta^2 = 0$ because of course one can emit *two* S-wave pions and still conserve parity.

In Fig. 7.7 is shown the π–π cross section results of Carmony and Van de

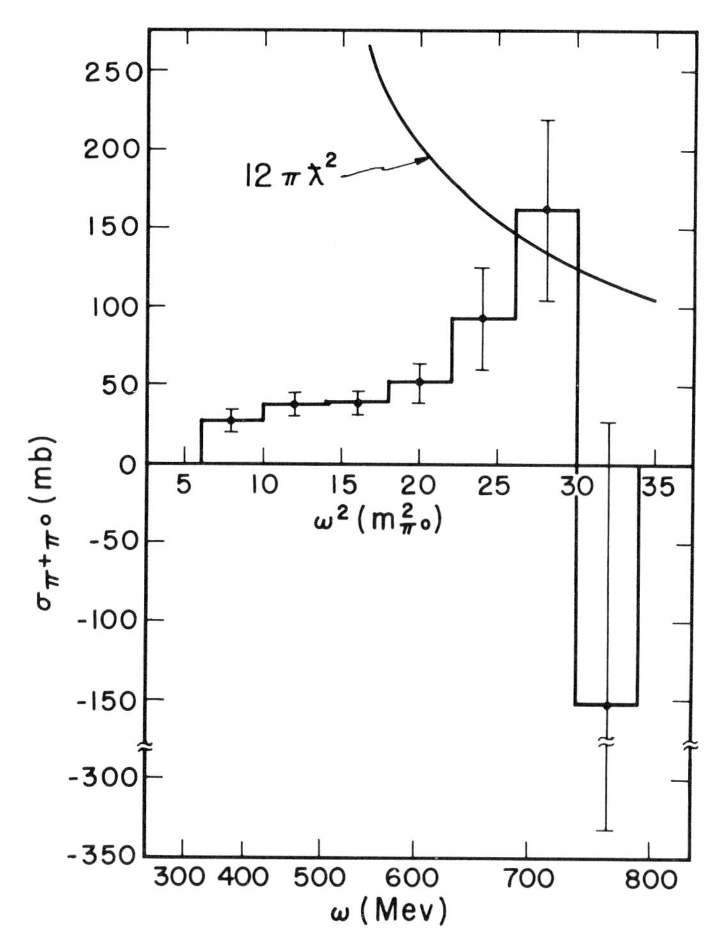

FIG. 7.7 The total cross sections for $\pi^+\pi^\circ \longrightarrow \pi^+\pi^\circ$ obtained by extrapolating reactions 7.1a and 7.1c to $\Delta^2 = -\mu^2$.

Walle [Ref. 10]. This data was obtained by extrapolating both reactions (7.1a) and (7.1c) to $\Delta^2 = -\mu^2$. This figure thus represents combined data for $\pi^+\pi^\circ$ and $\pi^-\pi^\circ$ scattering. The large statistical error is the result of the difficulties involved in extrapolating. A possible resonance is indicated at about 700 MeV with angular momentum $J = 1$. This angular momentum is suggested by the fact that the value of the π-π cross section at the peak is near $12\pi\lambda\!\!\!{}^{-2}$.

In order to try to obtain better statistical accuracy, Carmony and Van de Walle attempted to determine the π-π cross section directly from the data in the physical region. Formula (7.3) was used without extrapolating to $\Delta^2 = -\mu^2$. The results are shown in Fig. 7.8. Only data for which the recoil

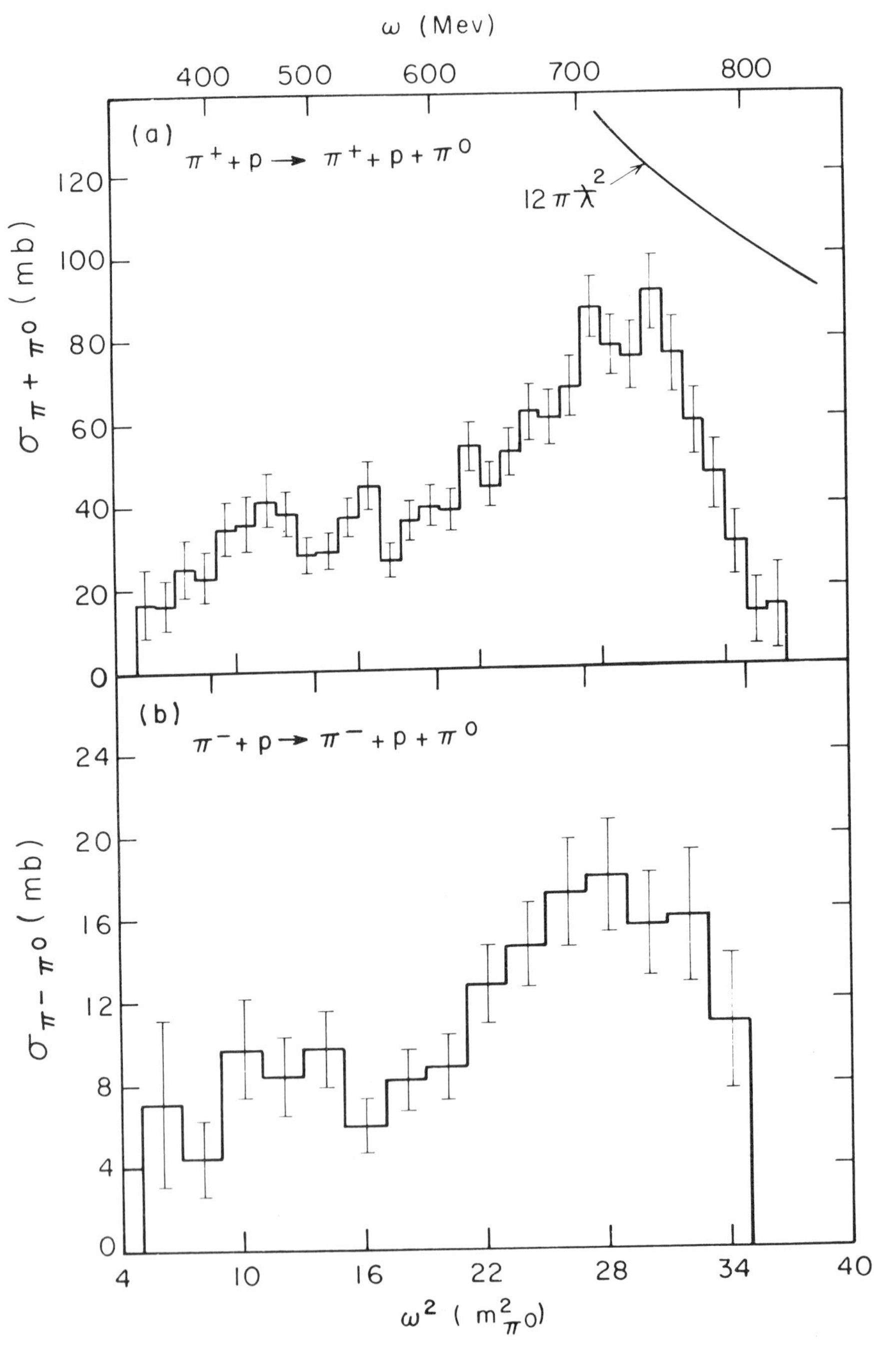

FIG. 7.8 The total pion-pion cross sections determined in the physical region from a) $\pi^+ p \longrightarrow \pi^+ p \pi^0$ and b) $\pi^- p \longrightarrow \pi^- p \pi^0$. Only events where the recoil proton momentum was < 400 MeV/c were used.

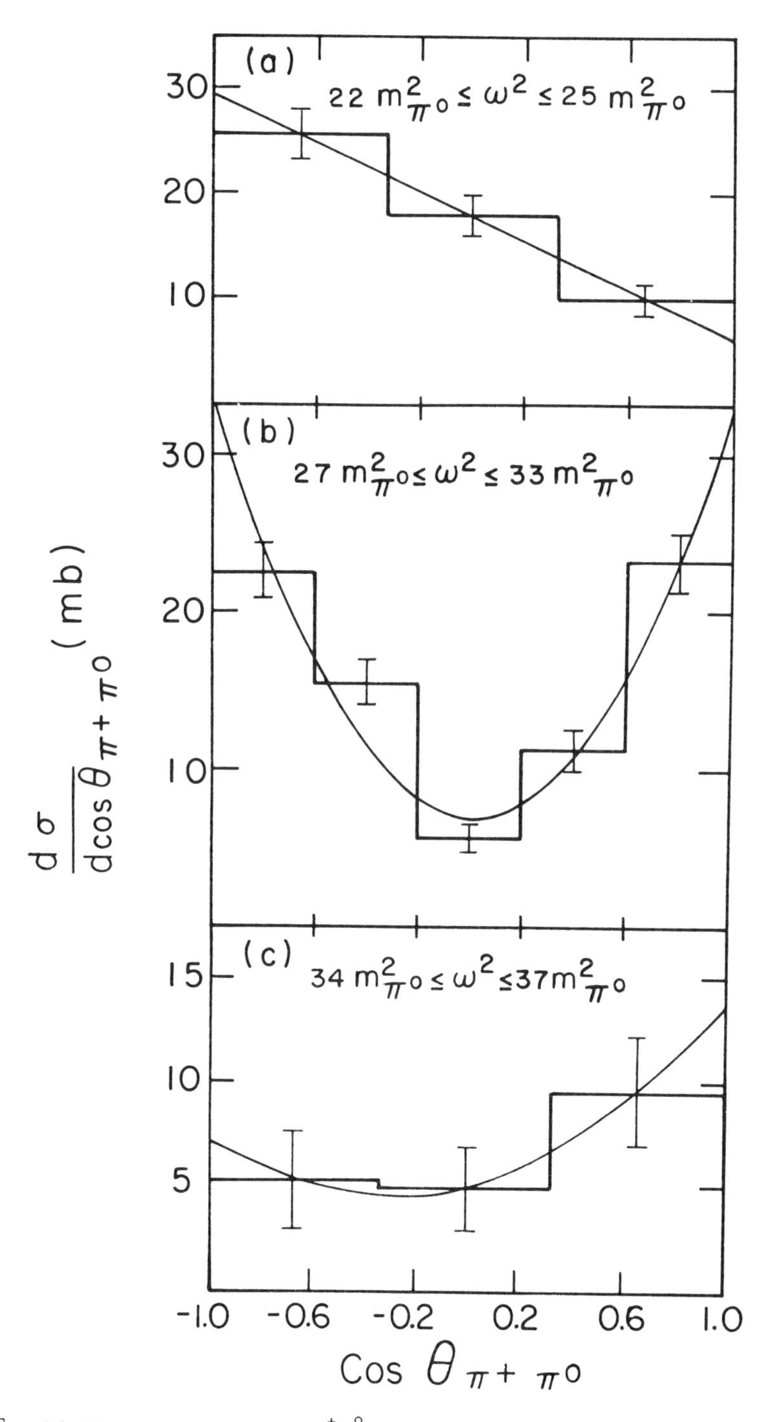

FIG. 7.9 Histograms showing the $\pi^+\pi^0$ differential cross section at three energies deduced from $\pi^+ p \longrightarrow \pi^+ p \pi^0$. Only events where the recoil proton momentum was <400 MeV/c were used.

proton momentum was < 400 MeV/c was included. This guarantees that the four momentum transfer was at least small. Both the π^+p and π^-p data indicate a peak in the π-π cross section near $28\, m_{\pi^\circ}^2$ (725 MeV). While the π^+p data shows a maximum π-π cross section near $12\,\pi\lambda\!\!\!^{-2}$ the π^-p data shows a much reduced π-π cross section. This indicates that for low momentum transfer one-pion exchange dominates the π^+p cross section but does not dominate the π^-p cross section.

In Fig. 7.9 is shown the π-π angular distribution from the π^+p data where again the recoil proton momentum was <400 MeV/c [Ref. 11]. Near the resonance energy it is clear that there is a dominant $\cos^2\theta$ term again pointing to a $J = 1$ resonance. Bose statistics then require that $T = 1$. This $J = 1$, $T = 1$ resonance at 725 MeV is just the ρ meson. At higher energies the ρ meson strongly affects the reactions (7.1a) to (7.1e) where it appears as a final state interaction.

7.5 The π-π Interaction. Low Energy

The extrapolation procedure discussed in the preceding section becomes difficult to apply at low energies. The π-π interaction must be deduced by other methods. There are several experiments which indicate that there is a strong attractive π-π interaction in the $T = 0$, $J = 0$ state at low energy. In addition, there is some evidence for a non-negligible interaction in the $T = 1$, $J = 1$ state. We will mention three experiments.

1. K^+ decay

The processes $K^+ \longrightarrow \pi^+ + \pi^+ + \pi^-$ and $K^+ \longrightarrow \pi^+ + \pi^\circ + \pi^\circ$ have been studied for some time to obtain information on a_0 and a_2, the $T = 0$ and $T = 2$ π-π scattering lengths.[1] Beg and de Celles, by taking account of the previously discussed P-wave $\pi\pi$ interaction, have shown that $a_0 \simeq 2\hbar/\mu c$ gives reasonable agreement with the $\pi^\pm$ spectra in the above decays [Ref. 12]. This corresponds to a 0 energy $T = 0$ π-π cross section of about 1,000 mb!

2. The ABC effect.

The ABC effect is the popular name given to the anomalous He^3 energy spectrum observed in the reaction

$$p + d \longrightarrow He^3 + 2\pi \tag{7.6}$$

What is observed is an enhancement at the upper end of the He^3 momentum spectrum. This effect does not appear in the spectrum of H^3 in the reaction

$$p + d \longrightarrow H^3 + \pi^+ + \pi^\circ \tag{7.7}$$

This indicates that the anomaly occurs only when the final pions are in the $T = 0$ state. The $T = 0$ part of reaction (7.6) is shown in Fig. 7.10

[1] The scattering length a may be simply defined by $\sigma_{\pi\pi} = 4\pi a^2$ at 0 kinetic energy.

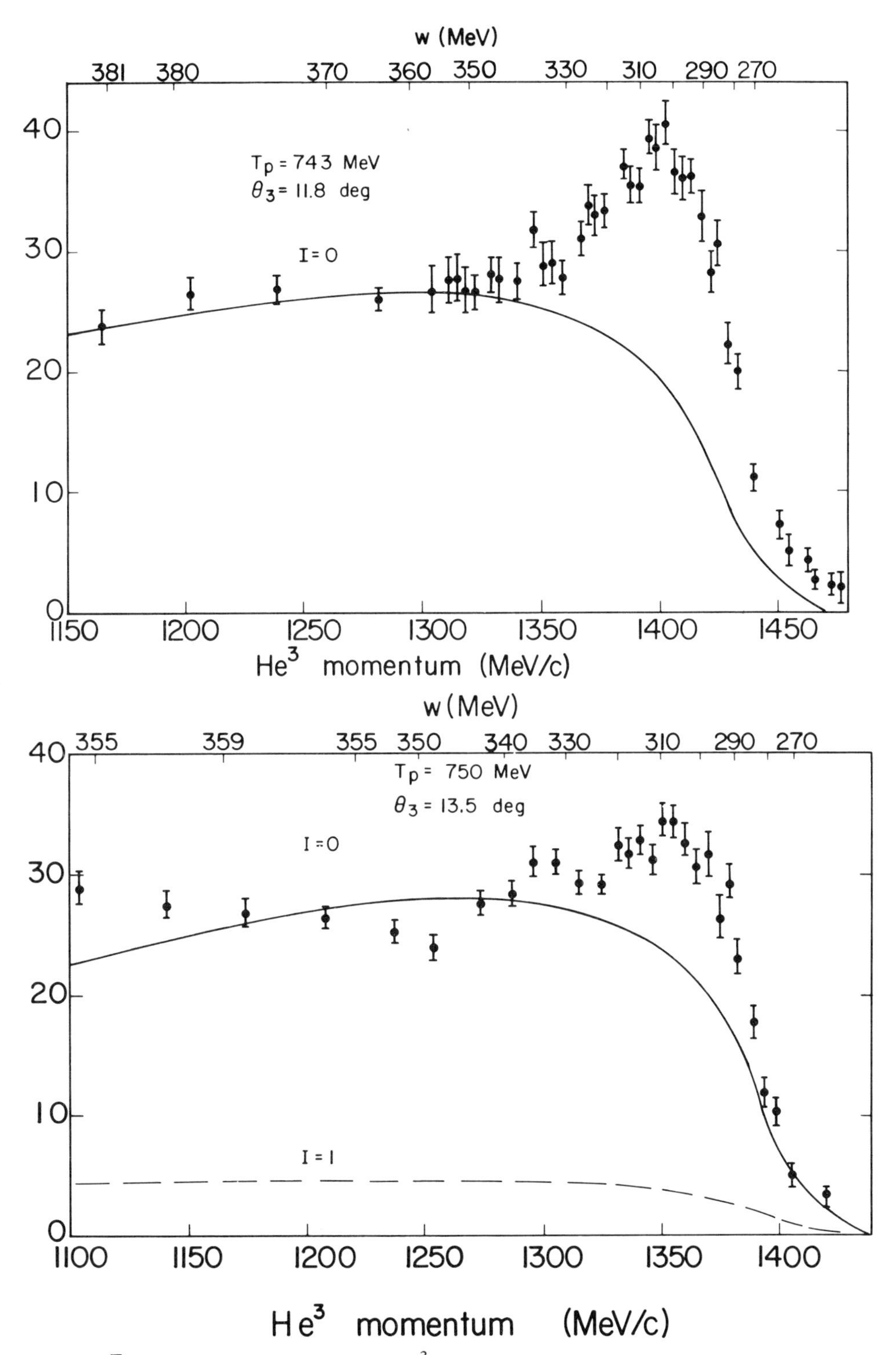

FIG. 7.10 The $T = 0$ part of the He^3 spectra at 11.8 and 13.5 deg. The curves are invariant phase space with the experimental resolution folded in. They are normalized to the low momentum points.

[Ref. 13]. The anomaly also disappears at wider laboratory angles indicating that it is a kinematic effect rather than a particle or resonance. In the region of the enhancement the two π's have minimum relative energy. These results have been interpreted as being due to a strong $T = 0$ $J = 0$ $\pi\pi$ interaction at 0 energy [Ref. 14]. A value $(2 \pm 1)\hbar/\mu c$ is deduced for a_0 [Ref. 13].

3. K_{e4} decay.

The process $K^+ \longrightarrow \pi^+\pi^- e^+\nu$ is ideal for investigating the π-π interaction. The only strongly interacting particles in the final state are the two π's. Unfortunately the branching ratio for this mode of decay is only 4×10^{-5}. The π-π interaction can be determined by measuring certain asymmetries in the final state. Based on 69 events observed in a propane bubble chamber a value of $(1 \pm 1)\hbar/\mu c$ is deduced for a_0 [Ref. 15]. The results of this experiment also indicate some interaction in the $T = 1, J = 1$ state.

Hamilton, Menotti, Oades, and Vick have calculated the $T = 0, J = 0$ π-π phase shift by using the experimental π-N phase shifts at low energy and partial wave dispersion relations derived from the Mandelstam representation [Ref. 16]. Using the solutions of the Chew-Mandelstam equations [Ref. 17] for π-π scattering they derive a value $a = (1.3 \pm .4)\hbar/\mu c$.

While none of the above three experiments determine the π-π scattering length a_0 with any precision, they all get values of (1 to 2) $\cdot$ $\hbar/\mu c$. This is corroborated by the above mentioned partial wave dispersion calculation.

7.6 The Isobar Model

The phenomenological model, due to Lindenbaum and Sternheimer, has been very useful in the description of π-N inelastic scattering [Ref. 18]. It assumes that single meson production takes place via production and subsequent decay of the $T = \frac{3}{2}, j = \frac{3}{2}$ nucleon isobar which we write as N^*. This is just the resonance that occurs in π-N scattering at 195 MeV. This isobar has a mass of 1238 MeV and a width of 90 MeV. This interaction is shown schematically in Fig. 7.11. The incoming pion scatters off the incoming nucleon producing an isobar. The isobar later decays into a nucleon and a

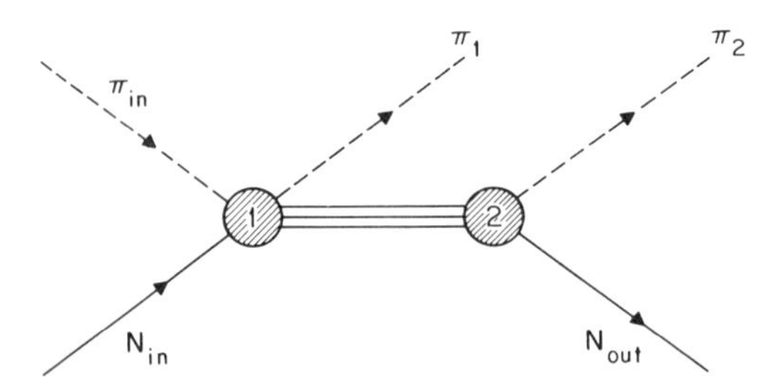

FIG. 7.11 Diagram of N^* (1238) exchange which is assumed to dominate in the isobar model calculation of $\pi N \longrightarrow \pi\pi N$.

pion. It is important to note that there is a second diagram in which the two final state pions π_1 and π_2 are interchanged. In order to facilitate calculations Lindenbaum and Sternheimer make three important approximations: (1) that interference between the two diagrams just mentioned can be neglected, (2) that the isobar is produced in an S-state, and (3) that the isobar decays isotropically. With these assumptions we can evidently write,

$$I_{\pi 1} = \left.\frac{d^2\sigma}{dT_\pi d\Omega_\pi}\right|_1 = a\,\sigma(m_I)\,F\frac{dm_I}{dT_\pi} \tag{7.8}$$

$$I_{\pi 2} = \left.\frac{d^2\sigma}{dT_\pi d\Omega_\pi}\right|_2 = b\int_{M_1}^{M_2} \sigma(m_I) F G_\pi\, dm_I \tag{7.9}$$

where,

T_π, Ω_π = kinetic energy and solid angle, respectively, of the final pions,
$I_{\pi 1}$ = the energy and angular distribution of pions produced at vertex 1 in Fig. 7.11,
$I_{\pi 2}$ = the energy and angular distribution of pions produced at vertex 2 in Fig. 7.11,
m_I = isobar mass,
$\sigma(m_I)$ = total $\pi^+ - p$ scattering cross section for an isobar of mass m_I,
F = the two body phase space factor for an isobar of mass m_I and the recoil pion,
G_π = the factor giving the energy spectrum of pions arising from the isobar of mass m_I, moving with velocity v_I,
$\frac{dm_I}{dT_\pi}$ = the range dm_I per unit range dT_π,
M_1, M_2 = lower and upper limits respectively of m_I allowed by energy and momentum conservation.

The constants a and b are determined from the total inelastic cross section and the condition,

$$\int_0^{T_{\pi\max}} I_{\pi 1}\, dT_\pi d\Omega_\pi = \int_0^{T_{\pi\max}} I_{\pi 2}\, dT_\pi d\Omega_\pi \tag{7.10}$$

This condition arises because there is always one pion produced at each vertex.

For isotropic decay of the isobar we have

$$\begin{aligned} G_\pi &= 1/(T_{\pi\max} - T_{\pi\min}), \; T_{\pi\min} < T_\pi < T_{\pi\max} \\ G_\pi &= 0, \; T_\pi < T_{\pi\min} \text{ and } T_\pi > T_{\pi\max} \end{aligned} \tag{7.11}$$

where $T_{\pi\min}$ and $T_{\pi\max}$ are respectively the minimum and maximum kinetic energies of the pions arising from the decay of an isobar with mass m_I and velocity v_I in the barycentric system.

In order to calculate the spectrum of π's of a given charge, it is necessary to take a linear combination of $I_{\pi 1}$ and $I_{\pi 2}$. The coefficients involved depend on two parameters: the ratio $\sigma(T = \frac{1}{2})_{\text{inel}}/\sigma(T = \frac{3}{2})_{\text{inel}}$ and the phase difference between the matrix element for pion production in the $T = \frac{1}{2}$ and $T = \frac{3}{2}$ state [Ref. 18].

In a similar way the recoil nucleon spectrum can be calculated. Namely,

$$I_n = \frac{d^2\sigma}{dT_n\, d\Omega_n} = b' \int_{M_1}^{M_2} \sigma(m_I) F G_n\, dm_I \tag{7.12}$$

G_n = the factor giving the energy distribution of the nucleons arising from the decay of an isobar of mass m_I, with velocity v_I. For an energy spectrum of nucleons integrated over all angles, $\frac{d\sigma}{dT_n}$, G_n is given by,

$$\begin{aligned} G_n &= 1/(T_{n\max} - T_{n\min}),\ T_{n\min} < T_n < T_{n\max} \\ G_n &= 0\ T_n < T_{n\min} \text{ and } T_n > T_{n\max} \end{aligned} \tag{7.13}$$

where $T_{n\min}$ and $T_{n\max}$ are, respectively, the minimum and maximum possible kinetic energies of the nucleons from the isobar decay.

The constant b' is determined by the condition,

$$\int I_{\pi 1}\, dT_\pi\, d\Omega_\pi = \int I_n\, dT_n\, d\Omega_n \tag{7.14}$$

similar to eq. (7.10). From I_n one can then determine the dipion mass spectrum.

Bergia, Bonsignori and Stanghellini extended the model to include the interference between the diagram of Fig. 7.11 and the diagram where π_1 and π_2 are interchanged [Ref. 19]. This is done by adding matrix elements directly rather than rates as was done by Lindenbaum and Sternheimer. Olsson and Yodh extended this work still further by including (1) the *P*-wave decay of the isobar and (2) the requirements of Bose statistics [Ref. 20].

7.7 π-N Inelastic Scattering. Discussion

First the general features mentioned at the beginning of the chapter will be discussed. Then the various models that have been proposed will be applied.

The smallness of the inelastic cross sections at low energies can be understood in terms of the diagram of Fig. 7.6. At low energies this diagram can be expected to dominate. Because the pion is pseudoscalar it must be emitted in a P state at vertex A. At low energies there is only a small amount of phase space available for emitting a P wave pion. A glance at reactions 7.1a and 7.1b show that the final pions must be emitted in isospin states $T = 1$ and 2 only. On the other hand, the pions in reactions 7.1d and 7.1e can be emitted in the $T = 0$ isospin state. From the discussion of Section 7.3 we expect a strong π-π interaction in this state at low energy. This enhances the interaction at vertex B in Fig. 7.6. Thus, we expect that reactions 7.1d and 7.1e will be larger than 7.1a,b and c in the energy region where the relative energy of the π's is small. Thus, the $\pi^- p$ inelastic cross section should be larger than the $\pi^+ p$ inelastic cross section at low energies. Fig. 7.2 shows that this is true.

From Table 7.2 the cross section for reaction 7.1e would be expected to be higher than for reaction 7.1d. However, as shown in Fig. 7.4, this is not

TABLE 7.2

Clebsch-Gordon coefficients for decomposition of π-π states into isospin states

	I_3	$I = 0$	$I = 1$	$I = 2$
$\pi^\circ\pi^\circ$	0	$\sqrt{\frac{2}{3}}$	0	$\sqrt{\frac{1}{3}}$
$\pi^+\pi^-$	0	$\sqrt{\frac{1}{6}}$	$\sqrt{\frac{1}{2}}$	$\sqrt{\frac{1}{3}}$
$\pi^+\pi^\circ$	+1	0	$\sqrt{\frac{1}{2}}$	$\sqrt{\frac{1}{2}}$
$\pi^-\pi^\circ$	−1	0	$\sqrt{\frac{1}{2}}$	$\sqrt{\frac{1}{2}}$
$\pi^+\pi^+$	+2	0	0	1
$\pi^-\pi^-$	−2	0	0	1

the case. Thus, the π-π interaction is not the whole story. The other effect that must be considered is the isobar effect discussed in section 7.6. From eqs. 1.12 it is seen that the isobar model favors reaction 7.1d. It is clear then that any attempt to understand single meson production must consider both the π-π interaction and the isobar effect.

Since the π-π interaction at low energies is weak in the $I = 1$ and 2 states, it should be possible to describe reactions 7.1a,b and c using only the isobar model. In Fig. 7.12 we compare experimental results for reactions 7.1a and b with the isobar calculation of Olsson and Yodh [Ref. 20]. The comparison is made at three energies 357, 600 and 820 MeV. The agreement is quite good. As one goes to higher energies (~1,000 MeV) the isobar model fails to give good agreement. This is because $Q(\pi^+, \pi^\circ)$ and $Q(\pi^-, \pi^\circ)$ distributions show peaks at the ρ mass reflecting the $T = 1, J = 1$, π-π resonance at 725 MeV. Even at low energies the isobar calculation fails to give the correct $Q(\pi^+, \pi^-)$ distribution due to the strong $T = 0, J = 0$ π-π interaction. At the same time

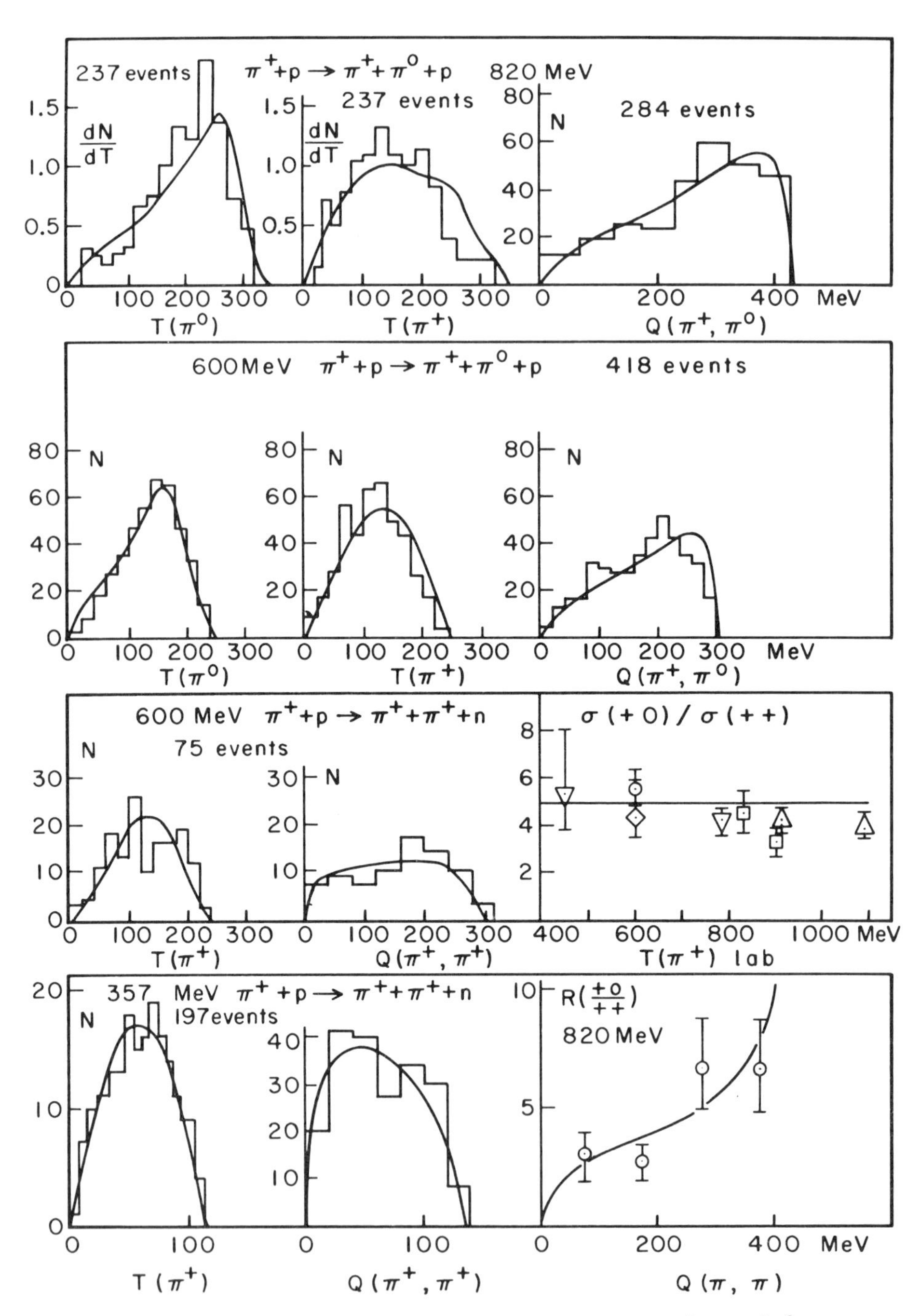

FIG. 7.12 Comparison of experimental data for the reactions $\pi^+ p \to \pi^+ \pi^\circ p$ and $\pi^+ p \to \pi^+ \pi^+ n$ with isobar model predictions.

it is clear the interference between the isobar effect and the π-π interaction is important. This can be seen from the experimental $M_{\pi^+\pi^-}$ and $M_{\pi^\circ\pi^\circ}$ distributions at 374 and 417 MeV shown in Fig. 7.13. If the π-π interaction were the only final state interaction, we would expect the $M_{\pi\pi}$ distributions to be enhanced at the low mass end. Fig. 7.13 shows that the opposite is the case.

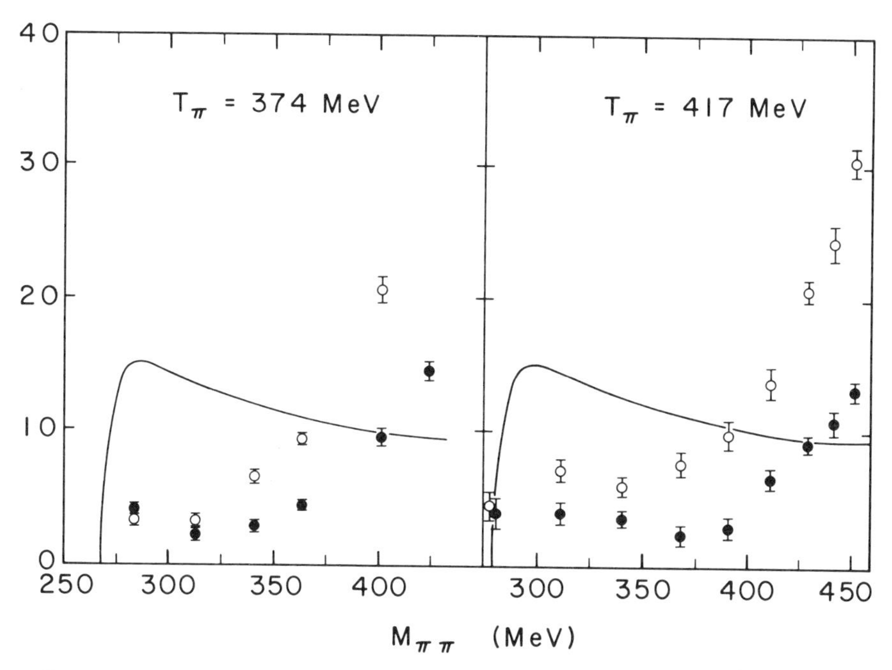

FIG. 7.13 The $M_{\pi\pi}$ energy distribution from the reactions $\pi^- p \longrightarrow \pi^+\pi^- n$ (○) and $\pi^- p \longrightarrow \pi^\circ\pi^\circ n$ (●). The curves are phase space.

Goebel and Schnitzer have calculated single meson production cross sections and angular distributions at low energies [Ref. 21]. They included both the π-π interaction and the isobar effect as final state interactions. The matrix elements were calculated using the static model of Chew and Low [Ref. 22]. There are three adjustable parameters in their calculation: a_0, a_1, a_2. These are the π-π scattering lengths in the Isospin states 0, 1, and 2, respectively. Here a_1 is defined as:

$$a_1 = \lim_{k \to 0} \sin \delta_1 / k^3 \tag{7.15}$$

where δ_1 is the $T = 1, J = 1$ phase shift and k is the center of mass momentum. Fitting the total $\pi^+ p$ inelastic cross section at 470 MeV and the π^+ angular distribution in reaction (7.1d) at 430 MeV, they find two acceptable sets of π–π scattering lengths. They are:

$$a_0 = 0.50, \quad a_1 = 0.07, \quad a_2 = 0.16$$

and

$$a_0 = 0.65, \quad a_1 = 0.07, \quad a_2 = -0.14$$

With these two sets of parameters Goebel and Schnitzer are able to obtain satisfactory fits to the cross sections for reactions 7.1a and 7.1e from threshold to about 500 MeV. However, they were able to obtain only qualitative agreement with the π^+ angular distribution in reaction 7.1d.

None of the calculations just described can be called definitive. The isobar calculation of Olsson and Yodh is phenomenological and omits the π–π interaction. The analysis of Goebel and Schnitzer is valid only at low energies where the static model is a good approximation. A complete dynamical calculation must include all the important final state interactions, namely the isobar effect, the $T = 0, J = 0$ π–π interaction and the ρ resonance. It must also include interferences between these final states.

7.8 The Reaction $\pi^- p \longrightarrow \eta n$

We conclude with a brief discussion of reaction 7.1f. As shown in Fig. 7.5 the cross section rises rapidly above threshold, peaks at about 650 MeV then gradually falls. The sharp rise is easily understood. The η has the same spin and parity as the pion. Thus, an S-wave final state is allowed. Near threshold then the cross section should be proportional to the momentum of the η in the barycentric system [Ref. 23]. The experimental results are consistent with this [Ref. 5,6]. The momentum of the η rises rapidly with pion lab energy near threshold and hence also the cross section. Above 650 MeV the angular distribution of η's becomes non-isotropic [Ref. 6] showing that higher partial waves are participating in the reaction. We note finally that the ηn final state does not seem to couple to the partial wave(s) responsible for the πN resonance at 900 MeV. The cross section for 7.1f shows no structure in this region.

REFERENCES

1. B. C. Barish, R. J. Kurz, V. Perez-Mendez, and J. Solomon, *Phys. Rev. 135,* 416 (1964).
2. J. C. Brisson, P. Falk-Vairant, J. P. Merlo, P. Sonderegger, R. Turlay, and G. Valladas in *Proceedings of the Aix-en-Provence International Conference on Elementary Particles,* CERN (Saclay, 1961) vol. 1, p. 45.
3. R. L. Rosenberg and L. D. Roper, *Lawrence Radiation Laboratory Report,* UCRL-14202, March 1965.
4. Data for Fig. 7.3 and 7.4 comes from the compilation of Ref. 3 and also Refs. 1 and 2.

5. Brown-Brandeis-Harvard-MIT-Padova Collaboration, *Phys. Rev. Letters 13,* 486 (1964).
6. W. B. Richards, C. B. Chiu, R. D. Eandi A. C. Helmholz, R. W. Kenney, B. J. Moyer, J. A. Poirier, R. J. Cence, V. Z. Peterson, N. K. Sehgal, and V. J. Stenger, *Phys. Rev. Letters 16,* 1221 (1966).
7. J. Kirz, J. Schwartz, and R. D. Tripp, *Phys. Rev. 130,* 2481 (1963).
8. R. J. Cence, V. Z. Peterson, V. J. Stenger, C. B. Chiu, R. D. Eandi, A. C. Helmholz, R. W. Kenney, B. J. Moyer, J. A. Poirier, and W. B. Richards, *Phys. Rev. Letters 19,* 1393 (1967).
9. G. F. Chew and F. E. Low, *Phys. Rev. 113,* 1640 (1959).
10. D. D. Carmony and R. T. Van de Walle, *Phys. Rev. 127,* 959 (1962).
11. D. D. Carmony and R. T. Van de Walle, *Phys. Rev. Letters 8,* 73 (1962).
12. M. A. Bég and P. C. De Celles, *Phys. Rev. Letters 8*, 46 (1962).
13. A. Abashian, N. E. Booth, K. M. Crowe, R. E. Hill, and E. H. Rogers, *Phys. Rev. 132,* 2296, 2305, 2309, 2314 (1963).
14. T. N. Truong, *Phys. Rev. Letters 6,* 308 (1961).
15. R. W. Birge, R. P. Ely, Jr., G. Gidal, G. E. Kalmus, A. Kernan, W. M. Powell, U. Camerini, D. Cline, W. F. Fry, J. G. Gaidos, D. Murphree, and C. T. Murphy, *Phys. Rev. 139,* B1600 (1965).
16. J. Hamilton, P. Menotti, G. C. Oades, and L. L. J. Vick, *Phys. Rev. 128,* 1881 (1962).
17. G. F. Chew, S. Mandelstam, *Phys. Rev. 119,* 467 (1960); G. F. Chew, S. Mandelstam, and H. P. Noyes, *ibid 119,* 478 (1960).
18. R. M. Sternheimer and S. J. Lindenbaum, *Phys. Rev. 109,* 1723 (1958).
19. S. Bergia, F. Bonsignori, and A. Stanghellini, *Nuo. Cim. 15,* 1073 (1960).
20. M. Olsson and G. B. Yodh, *Phys. Rev. Letters 10,* 353 (1963).
21. C. J. Goebel and H. J. Schnitzer, *Phys. Rev. 123,* 1021 (1961); H. J. Schnitzer, *Phys. Rev. 125,* 1059 (1962).
22. G. F. Chew and F. E. Low, *Phys. Rev. 101,* 1570 (1955).
23. M. Gell-Mann and K. M. Watson, *Ann. Rev. Nucl. Sci. 4,* 219 (1954).

CHAPTER 8

Photoproduction

8.1 Introduction

Because of its relevance to the pion-nucleon interaction we include a brief chapter on photoproduction. The reactions that will be considered are:

$$\gamma + p \longrightarrow \pi^+ + n, \tag{8.1a}$$

$$\gamma + p \longrightarrow \pi^\circ + p. \tag{8.1b}$$

The $\gamma + n$ interaction has been studied by using a deuterium target and taking advantage of the "spectator" process. These results will not be discussed here.

8.2 Experimental Methods

High Energy gamma ray beams are made by allowing electrons from an accelerator to strike a thin high-Z radiator. The electrons may have energies anywhere from 300 MeV to 20 GeV. The radiator is typically a few thousandths of an inch of tantalum. The gamma ray beam produced by the electrons passing through the radiator will have an energy spectrum characteristic of bremsstrahlung. This is shown for 300 MeV electrons in Fig. 8.1. The gamma ray beam will have an angular divergence $\sim m_e c^2 / E_e$ where E_e is the energy of the electrons impinging on the radiator. For 300 MeV electrons this angle is about 0.1 deg.

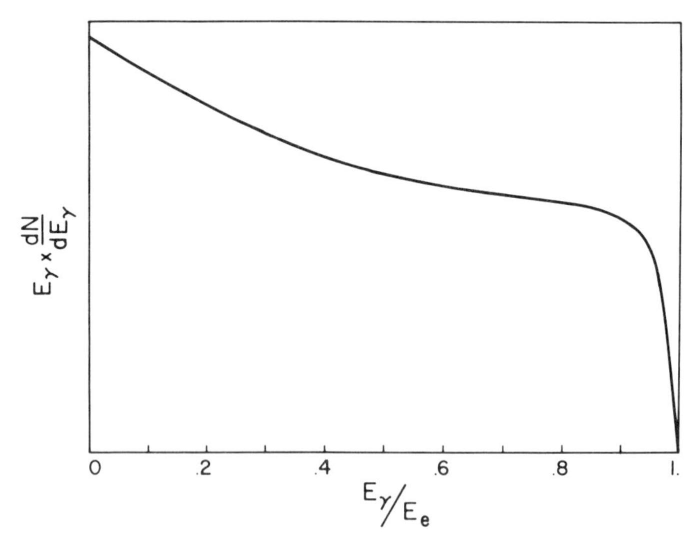

FIG. 8.1 Energy spectrum of gamma rays due to bremsstrahlung of 300 MeV electrons in a thin radiator. Note that the ordinate is $E_\gamma \times \dfrac{dN}{dE_\gamma}$ where $\dfrac{dN}{dE_\gamma}$ is the number of gammas per unit energy interval.

The final state in reactions 8.1a and b contain only two particles. Thus, a measurement of the angle of scattering and energy of one of the final particles determines the energy of the incoming gamma ray. It is necessary to assume that no more than one meson is produced in the reaction. This will be true only when the gamma ray energy deduced from reactions 8.1a and b is near the upper end of the spectrum of Fig. 8.1. This is because double meson production implies a higher gamma ray energy than single meson production for a given energy and scattering angle of the observed π^+ or proton. Thus, it is necessary that the energy implied by double meson production be beyond the upper limit of the gamma ray energy spectrum.

In a typical experiment a counter telescope and/or magnets are set up to measure the energy distribution of the π^+ (reaction 8.1a) or the proton (reaction 8.1b) at each of several angles. A differential ionization counter is included to distinguish between pions and protons. To determine the number of gammas passing through the target a thick walled ionization chamber is placed in the beam [Ref. 1]. It has been shown that the total ionization detected in such a chamber is proportional to the total energy of the gammas passing through it [Ref. 2]. The shape of the gamma ray spectrum (Fig. 8.1) is known from quantum electrodynamics and hence the number of gammas at each energy passing through the target is known.

Shown in Figs. 8.2 and 8.3 are the total cross sections for reactions (8.1a) and (8.1b) from 0 to 1.4 GeV [Ref's. 3, 4]. We notice immediately that there are peaks in these cross sections at energies *100 to 150 MeV higher than the corresponding peaks in π-N scattering.* Let us equate the total energy in the barycentric system for reactions (8.1) and for $\pi N \longrightarrow \pi N$:

$$E_B{}^2 = (M + \mu + T_\pi)^2 - \{(T_\pi + \mu)^2 - \mu^2\} = (M + \omega_L)^2 - \omega_L{}^2,$$

where ω_L is the lab energy of the gamma ray.
This simplifies to

$$\omega_L = T_\pi + \mu\left(1 + \frac{\mu}{2M}\right). \tag{8.2}$$

The term $\mu\left(1 + \frac{\mu}{2M}\right) = 150$ MeV for charged π's. Thus, the barycentric energies are the same if the γ ray bombarding energy is 150 MeV higher than the pion in the corresponding π-N scattering event. However, a study of Figs. 3.4, 3.5, and 8.2 shows that the $\gamma p \longrightarrow \pi^+ n$ peaks are all about 100 MeV rather than 150 MeV higher than the corresponding πN peaks. The reason for this will be given later.

The first peak at about 300 MeV is evidently to be identified with the $T = \frac{3}{2}$, $J = \frac{3}{2}$ isobar. From eqs. (1.12) we can predict that the ratio $\sigma(\gamma p \longrightarrow \pi^\circ p)/\sigma(\gamma p \longrightarrow \pi^+ n)$ due to the resonant state should be 2:1. *This is not observed.* The ratio is in fact about 6:5. To understand this, reactions (8.1) must be expanded into a sum of multipoles.

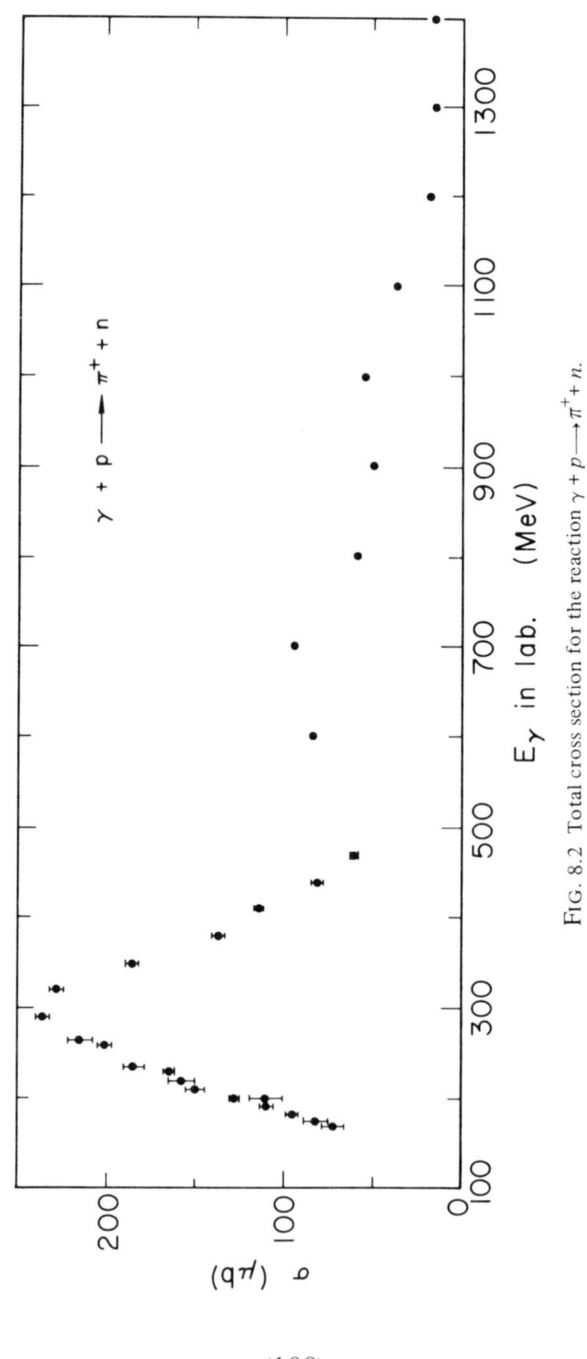

FIG. 8.2 Total cross section for the reaction $\gamma + p \longrightarrow \pi^+ + n$.

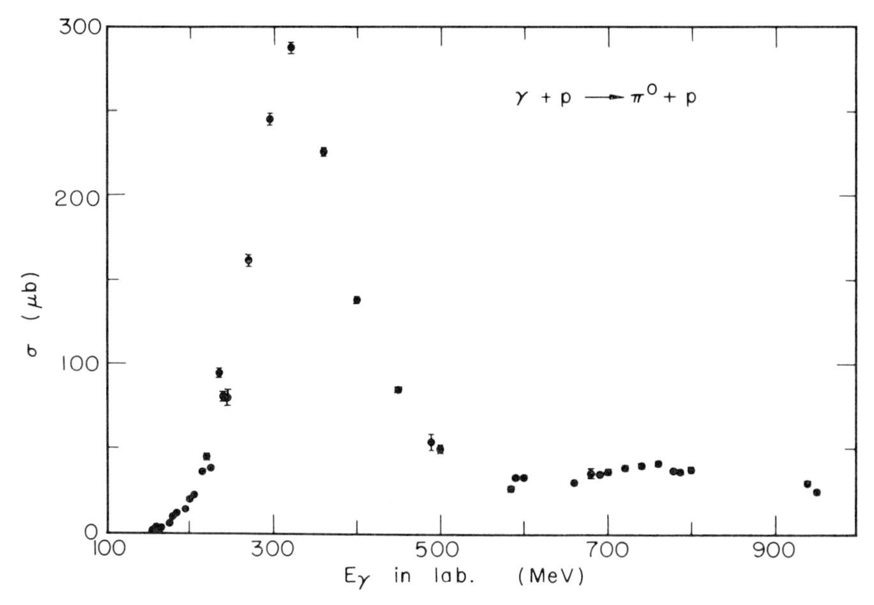

FIG. 8.3 Total cross section for the reaction $\gamma + p \longrightarrow \pi^{\circ} + p$.

The expansion of the radiation field into multipoles is an expansion into states of definite angular momentum and parity. Because the gamma ray has intrinsic spin 1, the total angular momentum j and the orbital angular momentum ℓ of a multipole are related by $j = \ell, \ell \pm 1$. Thus, a gamma ray described by a multipole of total angular momentum j can have either even or odd parity. The final states resulting from the electric and magnetic multipoles with $j = 1$ and 2 are shown in Table 8.1. It is assumed that the initial proton is an $S_{1/2}$ state.

The multipole which communicates with the $J = \frac{3}{2}$ resonance is evidently magnetic dipole ($j = 1$, even parity). The reason for the 6:5 ratio mentioned above is that the $\pi^+ n$ state has an appreciable $S_{1/2}$ state interaction which is greatly reduced in the $\pi^{\circ} p$ state. The $S_{1/2}$ state results from electric dipole absorption. If we assume, following Fermi, electric dipole absorption is proportional to the *static* dipole moment of the final state, then the ratio of reac-

TABLE 8.1

Multipoles with $j = 1$ and 2

Multipole	j	parity	final states (πp)
E_1	1	odd	$S_{1/2}$, $D_{3/2}$
M_1	1	even	$P_{1/2}$ $P_{3/2}$
E_2	2	even	$P_{3/2}$ $F_{5/2}$
M_2	2	odd	$D_{3/2}$ $D_{5/2}$

tions (8.1) would be $\sigma(\pi^+ n):\sigma(\pi^\circ p) = 1:\frac{\mu}{m}$. Near threshold this ratio is observed to be nearly correct as can be seen from Figs. 8.2 and 8.3. At the $T = \frac{3}{2}$, $J = \frac{3}{2}$ resonance reaction (8.1a) is a mixture of electric and magnetic dipole absorption while (8.1b) is magnetic dipole only. The 6:5 ratio observed at 300 MeV implies that E_1 absorption in reaction (8.1a) is comparable to M_1 absorption in reaction (8.1b).

The cross sections for reactions (8.1) show a second peak at 700 MeV for $\gamma p \rightarrow \pi^+ n$ and about 750 MeV for $\gamma p \longrightarrow \pi^\circ p$. This corresponds to the second π-N resonance at 600 MeV. In contrast to the first resonance, the second resonance shows reaction (8.1a) larger than (8.1b) with a ratio $\sigma(\gamma p \longrightarrow \pi^+ n)$: $\sigma(\gamma p \longrightarrow \pi^\circ p) \cong 3:1$. The resonance is evidently in the final state $T = \frac{1}{2}$ in agreement with the results from π-N scattering. The deviation from the ratio 2:1 which is predicted by eqs. (1.12) is due to a $T = \frac{3}{2}$ contribution to the scattering which evidently interferes destructively in reaction 8.1a.

The effect of the third resonance shows up as a very small bump at 1,000 MeV in the total cross section for $\gamma p \longrightarrow \pi^+ n$ (Fig. 8.2). It is not clear why this bump should be so small.

8.3 Angular Distributions and Polarization

Phenomenologically the scattering amplitude for photoproduction can be written as a sum of contributions from all possible multipole transitions characterized by multipole order j, total angular momentum J and parity p. Let α summarize these indices. Each multipole has a complex amplitude given by

$$M_\alpha e^{i\delta_\alpha}$$

where M_α and δ_α are real numbers. Then the differential cross section may be written:

$$\frac{d\sigma}{d\Omega} = \sum_{\alpha} M_\alpha^2 f_\alpha(x) + \sum_{\alpha<\beta} M_\alpha M_\beta \cos(\delta_\alpha - \delta_\beta) f_{\alpha\beta}(x) \tag{8.3}$$

and the polarization $\vec{P}$ of the recoil nucleon as

$$P\left(\frac{d\sigma}{d\Omega}\right) = \vec{n}_\perp \sum_{\alpha>\beta} M_\alpha M_\beta \sin(\delta_\beta - \delta_\alpha) g_{\alpha\beta}(x) \sin\theta \tag{8.4}$$

where $x = \cos\theta$, $\theta =$ the pion angle in the barycentric system, and $\vec{n}_\perp = (\vec{q} \times \vec{k})/|\vec{q} \times \vec{k}|$ where $\vec{q}$ and $\vec{k}$ are respectively the momenta of the incoming photon and the outgoing pion in the barycentric system. The functions $f_\alpha(x)$, $f_{\alpha\beta}(x)$, $g_{\alpha\beta}(x)$ are polynomials in x determined by the angular momentum and parity associated with the multipoles α and β. Table B.1 in Appendix B displays these functions through $J = \frac{5}{2}$ following Peierls [Ref. 5]. For a derivation see Hayakawa, Kawaguchi, and Minami [Ref. 6]. Hoff has derived distributions using polarized photons [Ref. 7]. Watson has shown that at low energies, where there is no inelastic scattering, δ_α is identical with the $\pi N \longrightarrow \pi N$ scattering phase shift in the state with the same J, P and T [Ref. 8].

Note that there is an equivalent to the Minami ambiguity for photoproduction. A study of Table B.1 shows that if the amplitudes for electric and magnetic multipoles of the same order are interchanged, the differential cross section remains unchanged while the polarization changes sign. If now the sign of all the phases δ_α are reversed, the polarization will again change sign as can be seen from eq. (8.4). Thus, if both the above changes are made, the differential cross sections and the polarization remain unchanged. As in π-N scattering, then, the total angular momentum of the final state can be determined from the angular distributions and the polarization but not the parity.

Strictly speaking, eqs. (8.3) and (8.4) are correct only for reaction (8.1b), $\gamma p \longrightarrow \pi^\circ p$. As emphasized by Moravcsik, all meson theories predict a term in the scattering amplitude for $\gamma p \longrightarrow \pi^+ n$ of the form [Ref. 9],

$$\frac{(\vec{\epsilon} \cdot \vec{k})\vec{\sigma} \cdot (\vec{q} - \vec{k})}{\omega k(1 - \beta \cos \theta)} \tag{8.5}$$

where $\vec{q}$ and $\vec{\epsilon}$ are the photon momentum and polarization, $\vec{k}$, ω are the momentum and energy of the meson, βc is the velocity of the pion, and $\vec{\sigma}$ is the nucleon spin. This "retardation" term in the scattering amplitude corresponds to the direct ejection by a photon of a virtual meson from the meson cloud surrounding the nucleon. It is analogous to the photoelectric effect in atomic physics. The Feynman diagram for this process is shown in Fig. 8.4. It does not occur for neutral meson production. Due to the denominator in expression (8.5), this term will contribute to all multipoles in eqs. (8.3) and (8.4). Its effect is most noticeable at small angles [Ref. 10]. Furthermore, it will affect the total cross section since it interferes with all multipoles.

It has been shown by Wetherell that interference between the retardation term and the resonant multipoles is important [Ref. 11]. The effect is to reduce by about 50 MeV the energy at which the total cross section peaks in $\gamma p \longrightarrow \pi^+ n$. This agrees with our earlier observation from the experimental results shown in Fig. 8.2. A more refined calculation has been made by Ph. Salin [Ref. 12].

It is convenient to express the experimental results for the differential cross sections by a power series in $\cos \theta$.

$$\frac{d\sigma}{d\Omega} = \sum_n A_n \cos^n \theta . \tag{8.6}$$

The coefficients for Reaction (8.1a), written A_n^+, are shown in Fig. 8.5 [Ref. 3] and those for reaction 8.1b, written A_n°, are shown in Fig. 8.6

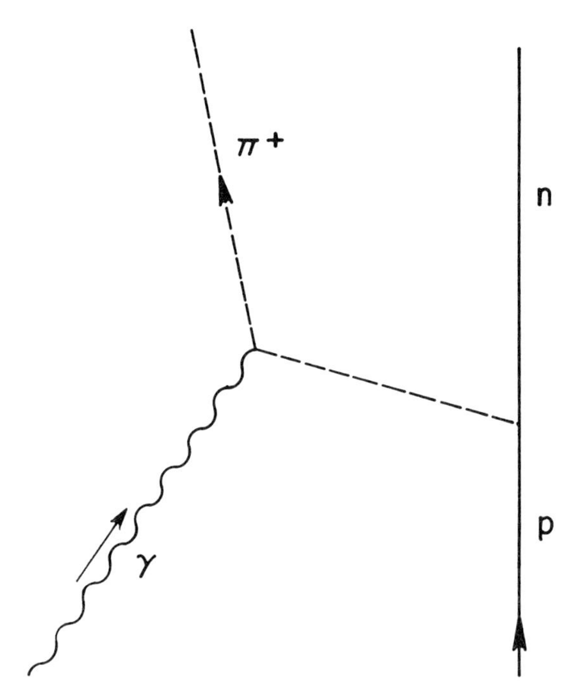

FIG. 8.4 Diagram for photo-ejection of a meson from the meson cloud of the nucleon.

[Ref. 4]. Because of the retardation term in $\gamma p \longrightarrow \pi^+ n$ the coefficients A_n^+ are only interpretable in terms of multipoles in an approximate way.

From Figs. 8.5 and 8.6 it is seen that the angular distributions for both reactions (8.1a) and (8.1b) are approximately $\propto(5 - 3\cos^2\theta)$ near the first resonance at 300 MeV. Table B.1 shows that only states with $J = \frac{3}{2}$ have this angular distribution. This agrees with the results obtained from the phase shift analysis of the first πN resonance.

Near the second resonance (700 MeV for Reaction (8.1a) and 750 MeV for Reaction (8.1b)), the angular distributions are quite different. The $\pi^+ n$ distribution is approximately $\propto(2 + 2\cos\theta - \cos^3\theta)$. This does not correspond to any single multipole. It evidently reflects the effect of the retardation term. The $\pi^\circ p$ distribution is again approximately $\propto(5 - 3\cos^2\theta)$. This would indicate $J = \frac{3}{2}$ for the second resonance. This conclusion must be accepted with caution, however. The total cross section results, Fig. 8.3, show that there is considerable non-resonant background. This must contribute to the angular distribution.

Attempts have been made to relate the parities of first and second resonances by measuring the polarization of the proton in (8.1b) at 90 deg. A study of Table 8.2 shows that only an interference between two states of opposite partiy can contribute. Fig. 8.7 shows the experimental results [Ref. 13]

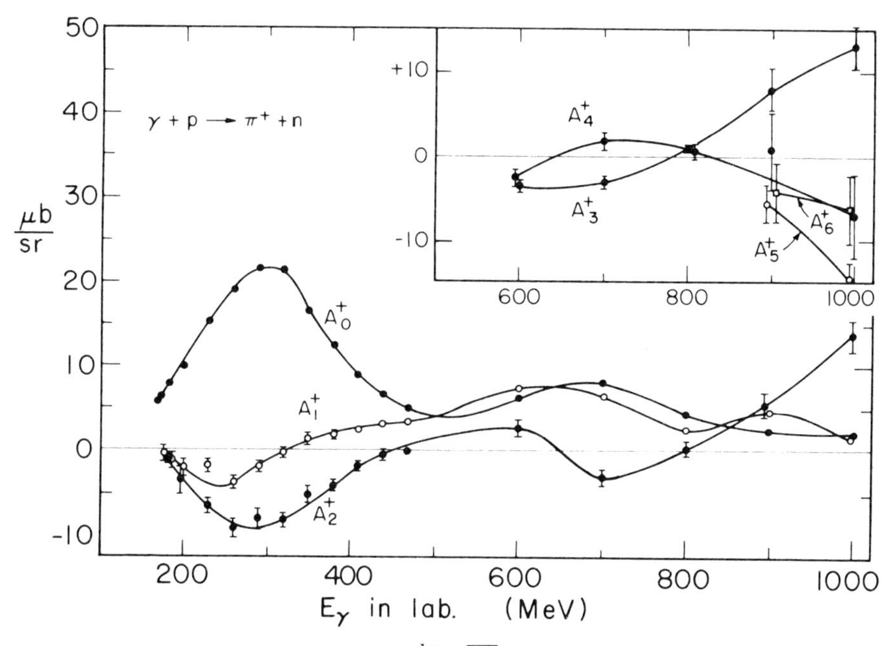

FIG. 8.5 Coefficients in the expansion $\frac{d\sigma}{d\Omega} = \sum A_n^+ \cos^n \theta$ for $\gamma p \longrightarrow \pi^+ n$. The curves were drawn by eye.

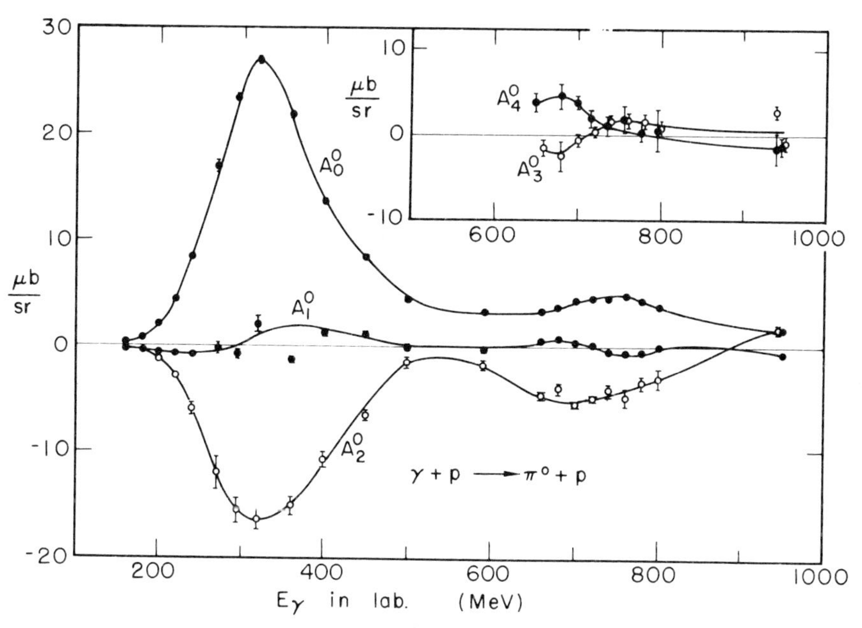

FIG. 8.6 Coefficients in the expansion $\frac{d\sigma}{d\Omega} = \sum A_n^\circ \cos^n \theta$ for $\gamma p \longrightarrow \pi^\circ p$. The curves were drawn by eye.

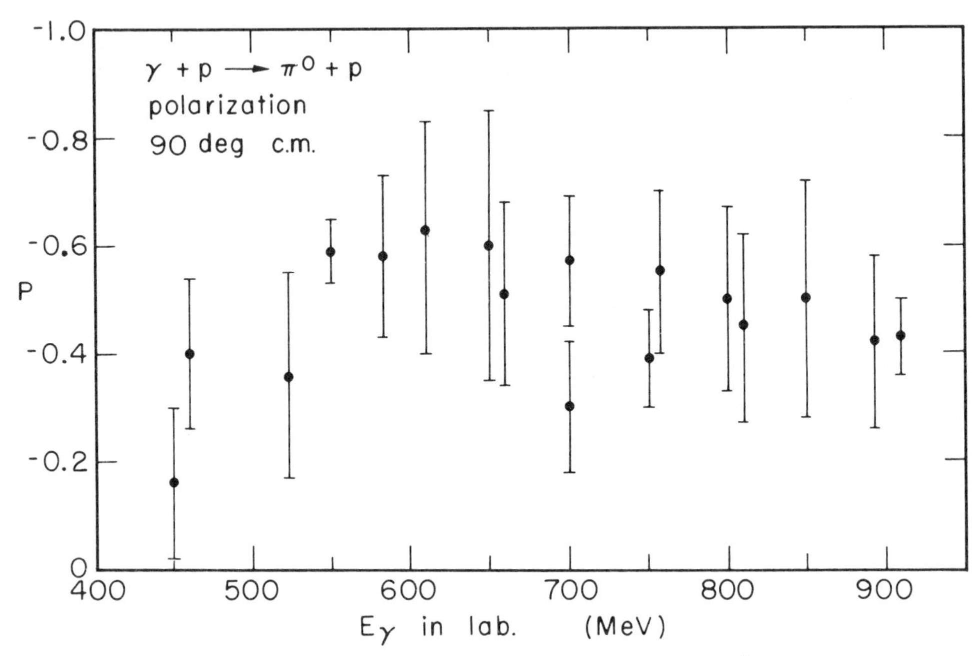

FIG. 8.7 Polarization of the recoil nucleon at 90 deg. cm. in $\gamma p \longrightarrow \pi^0 p$. Positive polarization is defined to be along $\vec{k} \times \vec{q}$. This is opposite to the convention used in eq. 8.4.

above the first resonance. It is striking that from about 450 MeV to 900 MeV the polarization is large and negative using the definition of $\vec{n}_\perp$ in the caption to Fig. 8.7. If it is assumed that this polarization results from an interference between the P_{33} resonance at 325 MeV and the second resonance at 750 MeV, these results imply that the first and second resonances have *opposite* parity [Ref. 14]. This would make the second resonance a $D_{3/2}$ state since the first is a $P_{3/2}$ state. If further the theorem of Watson [Ref. 8] is assumed to be valid at these energies, the polarization is correctly predicted to be negative. In expression (8.4) the state β is P_{33} and α is D_{13}. Hence, we expect $180 > \delta_\beta - \delta_\alpha > 90$ deg. at energies between the two resonances. According to the convention stated in the caption for Fig. 8.7, $\vec{P}$ is then negative. This argument assumes that between the two resonances only the multipoles responsible for the resonances are appreciable. However, the phase shift analyses have shown that in π-N scattering the amplitudes for several partial waves are large in this energy region (e.g., S_{11} and P_{11}). The argument for $D_{3/2}$ is thus weakened considerably.

REFERENCES

1. R. R. Wilson, *Nucl. Instr. 1*, 101 (1957).
2. W. Blocker, R. Kenney, and W. Panofsky, *Phys. Rev. 79*, 419 (1950).

3. Data for Figs. 8.2 and 8.5 come from: F. P. Dixon, "*Photoproduction of Positive Pions from Hydrogen in the 600 to 1,000 MeV Region*," California Institute of Technology (unpublished) (1960); M. Beneventano, G. Bernardini, D. Carlson-Lee, G. Stoppini, and L. Tau, Nuo. Cim. *4*, 323 (1956); R. L. Walker, J. G. Teasdale, V. Z. Peterson, and J. L. Vette, *Phys. Rev. 99*, 210 (1955); A. V. Tollestrup, J. C. Keck, R. M. Worlock, *Phys. Rev. 99*, 220 (1955).
4. Data for Figs. 8.3 and 8.6 come from: H. De Staebler, Jr., E. F. Erickson, A. C. Hearn, and C. Schaerf, *Phys. Rev. 140,* B336 (1965); K. Berkelman and J. A. Waggoner, *Phys. Rev. 117*, 1364 (1960); R. G. Vasil'kov, B. B. Govorkov, and V. I. Gol'danskii, *Soviet Physics JETP 10,* 7 (1960); J. I. Vette, *Phys. Rev. 111*, 622 (1958); L. J. Koester and F. E. Mills, *Phys. Rev. 105*, 1900 (1957) (corrected data received by private communication.).
5. R. F. Peierls, *Phys. Rev. 118*, 325 (1960), and errata, *Phys. Rev. 124*, 2051 (1961).
6. S. Hayakawa, M. Kawaguchi and S. Minami, *Suppl. Progr. Theor. Phys.* (Kyoto) No. *5*, 41 (1958).
7. G. T. Hoff, *Phys. Rev. 122*, 665 (1961).
8. K. Watson, *Phys. Rev. 95*, 228 (1954).
9. M. J. Moravcsik, *Phys. Rev. 104*, 1451 (1956).
10. E. A. Knapp, R. W. Kenney, and V. Perez-Mendez, *Phys. Rev. 114*, 605 (1959).
11. A. M. Wetherell, *Phys. Rev. 115*, 1722 (1959).
12. Ph. Salin, *Nuo. Cim. 28,* 1294 (1963).
13. Data for Fig. 8.7 comes from: J. O. Maloy, V. Z. Peterson, G. A. Salandin, F. Waldner, A. Manfredini, J. I. Friedman, and H. Kendall, *Phys. Rev. 139,* B733 (1965); R. Querzoli, G. Salvini and A. Silverman, *Nuo. Cim. 19*, 57 (1961); C. Mencuccini, R. Querzoli, and G. Salvini, *Aix-en-Provence Conf. 1,* 17 (1961); P. C. Stein, *Phys. Rev. Letters 2,* 473 (1959).
14. J. J. Sakurai, *Phys. Rev. Letters 1*, 258 (1958).

CHAPTER 9

Models

In this final chapter we will discuss some of the current theoretical ideas regarding the various resonances in π-N scattering. As the title implies there does not exist a *theory* which predicts the properties of the various resonances that are observed. In spite of the enormous effort expended, a comprehensive dynamical theory has so far eluded physicists. There are, however, models which describe some aspects of the π-N interaction.

9.1 The Chew-Low Model

We begin by describing the cut-off model of Chew and Low [Ref. 1]. This model successfully predicts the gross characteristics of π-N scattering at pion energies below a few hundred MeV. In particular it predicts the resonance in the $T = \frac{3}{2}$, $j = \frac{3}{2}$ state. This model has been described in detail in several texts [Ref. 2]. Therefore, only a brief outline will be given here.

The pseudovector interaction between pions and nucleons is assumed:

$$L_I = f^{(\circ)} \psi^\dagger [(\vec{\sigma} \cdot \vec{\nabla})(\vec{\tau} \cdot \vec{\phi})] \psi \tag{9.1}$$

where

L_I = the interaction term in the Lagrangian density
ψ = nucleon field operator
$\vec{\phi}$ = pion field operator, a vector in isotopic spin space
$\vec{\sigma}$ = Pauli spin operator
$\vec{\tau}$ = isotopic spin operator, a vector in isospin space
$f^{(\circ)}$ = rationalized but unrenormalized coupling constant

The $\vec{\nabla}$ operates on $\vec{\phi}$. This interaction is clearly not relativistically invariant. It uses only the large components of the spinors $\psi^\dagger$ and ψ.

It is further assumed that the nucleon is at rest. Then an interaction Hamiltonian is deduced,

$$\begin{gathered} H_I = \sum_m (V_m^{(\circ)} a_m + V_m^{(\circ)\dagger} a_m{}^\dagger) \\ V_m^{(\circ)} = i f^{(\circ)} (\vec{\sigma} \cdot \vec{k}/\sqrt{2\omega}) \tau_m \; v(k^2) \end{gathered} \tag{9.2}$$

where

$a_m^\dagger, a_m$ = respectively, creation and annihilation operators for single mesons,
$\vec{k}$ = momentum of the meson in units of μc,
$\omega = \sqrt{k^2 + 1}$ = total energy of the pion (mass $\mu = 1$),
m summarizes all the meson quantum numbers,
$v(k^2)$ = Fourier transform of the nucleon charge density

and

$$v(k^2) = \int e^{-i\vec{k}\cdot\vec{r}} \rho(\vec{r}) d\vec{r}$$

In order to ensure the convergence of the necessary integrations $v(k^2)$ must be cut off above some k_{max}. Thus, we assume

$$\begin{aligned} v(k^2) &= 1 \quad k < k_{max} \\ &= 0 \quad k > k_{max} \end{aligned} \tag{9.3}$$

There are two parameters in the model which must be determined from experiment: (1) k_{max} and (2) f which is the coupling constant $f^{(\circ)}$ after renormalization.

By virtue of the pseudovector interaction and the assumption that the nucleon remains rigidly at rest, the model predicts that all scattering will be in the P-state. Further, it says that:

$$\delta_{1\,1/2}(T = \tfrac{3}{2}) = \delta_{1\,3/2}(T = \tfrac{1}{2}) \tag{9.4}$$

using the notation of Chapter 2 where we write $\delta_{\ell j}$.

The absence of interaction in states of orbital angular momenta other than $\ell = 1$ is easily understood from general conservation laws. The Yukawa theory is based on the emission and absorption of mesons one at a time by the nucleon. Parity and angular momentum are conserved in each elementary process. It is clear then that the odd intrinsic parity and zero spin of the π meson means that there can be meson absorption or emission only when the orbital angular momentum is equal to one.

Finally use is made of a scattering equation derived by Low [Ref. 3]. To solve this equation it is assumed that only the two diagrams where a nucleon and a meson are exchanged are important. These are shown in Fig. 9.1. Finally, the following relation is derived for the $T = \frac{3}{2}, j = \frac{3}{2}$ phase shift

$$\frac{k^3}{\omega} \cot \delta_{1\,3/2}(T = \tfrac{3}{2}) = \frac{3}{4f^2}(1 - r\omega) \tag{9.5}$$

where f is the renormalized coupling constant. Here r is a constant whose value is approximately related by

$$r \cong f^2 \omega_{max} > 0 \tag{9.6}$$

where $\omega_{max} = \sqrt{k_{max}^2 + 1}$, k_{max} = cut off momentum. If the left side of equation (9.5) is plotted vs. ω a straight line should result. If the line is extrapolated to $\omega = 0$, the coupling constant f^2 can be determined. Since $r > 0$ the slope of the line should be negative. The point at which the line crosses the abscissa will be the energy at which there is a resonance in the $T = \frac{3}{2}$, $j = \frac{3}{2}$ state since $\delta_{1\,3/2}(T = \frac{3}{2}) = 90$ deg. Fig. 9.2 shows such a plot. The curves come from the energy dependent phase shift analyses, Refs. 16 and 17 of Chapter 6.

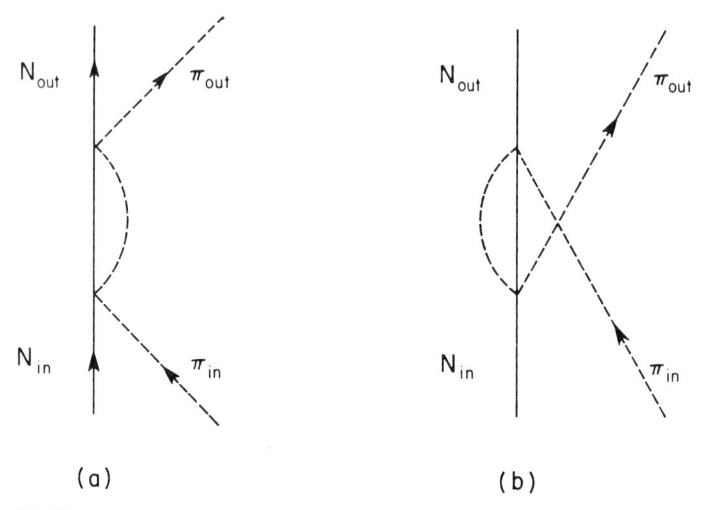

FIG. 9.1 The two diagrams that are dominant in the cut off model of Chew and Low.

In Fig. 9.2 ω is the total energy of the pion plus the kinetic energy of the nucleon. That is

$$\omega = \sqrt{k^2 + 1} + \frac{k^2}{2M} \tag{9.7}$$

This slightly different definition gives better agreement with relation (9.5). There are plausibility arguments based on relativistic dispersion theory that indicate nucleon recoil would produce the modification indicated by eq. (9.7).

The curves of Fig. 9.2 are not straight. But as $\omega \longrightarrow 1$ they appear to approach a straight line. If they are extrapolated thru the non-physical region $\omega < 1$ to $\omega = 0$, the coupling constant is found to be,

$$f^2 \cong .087$$

Also from Fig. 9.2,

$$\omega_{\text{res}} = 2.1$$

This corresponds to $T_\pi = 195$ MeV in the lab. The slope r appears to be $\sim\frac{1}{2}$. This makes $\omega_{\text{max}} \sim 6$. The two parameters r and f^2 of the cut off model are thus determined from Fig. 9.2.

Relations similar to (9.5) are predicted for the other P-wave phase shifts. However, for these phase shifts $r < 0$. This means that there are no resonances in these other partial waves. Experimentally the other phase shifts are small. Also, the S-wave phase shifts which are zero in this model are small experimentally.

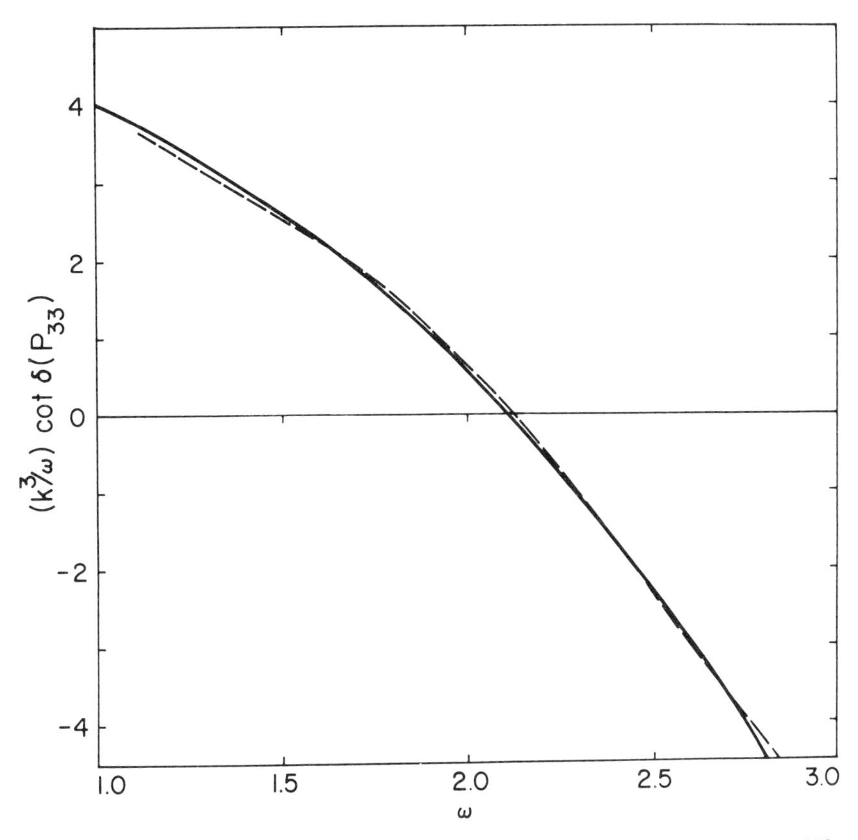

FIG. 9.2 Plot of eq. 9.5. The curves came from the energy dependent phase shift analyses, Refs. 16 and 17 of Chapter 6.

The Chew-Low model was a break-through for field theory. It was the first calculation involving a strong interaction which gave any significant agreement with experiment. By its specific assumptions it is limited to low energies. Furthermore, strange particles play no role in this model. It is clear that any model that is valid at higher energies must account for the copious production of these particles.

9.2 General Ideas Concerning the Higher Resonances

Because of its restrictive assumptions it is not possible to generalize the cut off model to the higher resonances. The properties of these resonances grouped according to isotopic spin and parity are summarized in Table 9.1. They exhibit certain patterns that are worth pointing out.

The most obvious observation is that resonances with higher mass have higher spins. Fig. 9.3 shows a plot of $(\text{mass})^2$ versus spin for the various resonances. Each line includes resonances with a single isotopic spin and parity.

TABLE 9.1

Properties of π-N Resonances

Mass	J	Parity	Isotopic spin
938	$\frac{1}{2}$	+	$\frac{1}{2}$
1,688	$\frac{5}{2}$	+	$\frac{1}{2}$
1,519	$\frac{3}{2}$	−	$\frac{1}{2}$
2,190	$\left(\frac{7}{2}\right)$	−	$\frac{1}{2}$
2,650	$\left(\frac{11}{2}\right)$	(−)	$\frac{1}{2}$
1,238	$\frac{3}{2}$	+	$\frac{3}{2}$
1,920	$\frac{7}{2}$	+	$\frac{3}{2}$
2,420	$\left(\frac{11}{2}\right)$	(+)	$\frac{3}{2}$
2,850	$\left(\frac{15}{2}\right)$	(+)	$\frac{3}{2}$
1,670	$\frac{1}{2}$	−	$\frac{3}{2}$

Values in parentheses have not been measured but are deduced from Fig. 9.3.

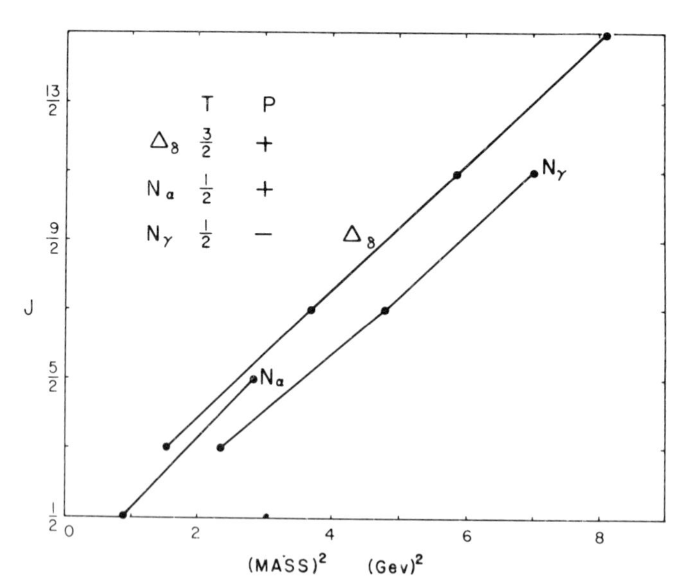

FIG. 9.3 Plot of spin vs. (mass)^2 of the various π-N resonances.

One resonance (1670 MeV) is not included. It does not seem to follow the pattern exhibited by the other resonances and in fact there is some doubt that it is a genuine resonance. If only the lines in Fig. 9.3 with more than two points are included then for

$T = \frac{1}{2}$ we have $j = \ell - \frac{1}{2}$ odd parity

and for $T = \frac{3}{2}$ we have $j = \ell + \frac{1}{2}$ even parity

It is remarkable that the points lie on such straight lines. These lines are in fact used to deduce the spins of 4 of the resonances listed in Table 9.1. The slope of these lines is easily understood. Assume that as each angular momentum comes within the range of the π-N force a resonance occurs. Then from eq. (1.9)

$$k^2R^2 \approx \ell(\ell + 1)$$

where k = the wave number in the barycentric system and ℓ = the orbital angular momentum. Differentiating then,

$$\frac{\Delta\ell}{\Delta k^2} \approx \frac{R^2}{2\ell + 1} \tag{9.8}$$

Since $\Delta m^2 \cong 4\Delta k^2$ ($\hbar = c = 1$) to within a factor of 2,

$$\frac{\Delta\ell}{\Delta m^2} \approx \frac{1}{4}\,\frac{R^2}{2\ell + 1} \tag{9.9}$$

If we take $R = \frac{1}{2}$ (pion Compton wave length) = 0.7 Fermis and $\ell = 1$ then

$$\frac{\Delta\ell}{\Delta m^2} \approx 1\,(\mathrm{GeV})^{-2} \tag{9.10}$$

in agreement with Fig. 9.3. Thus, the slope of the lines in Fig. 9.3 is merely a consequence of the finite range of the π-N force. The argument leading to (9.10) is only an order of magnitude. Thus, the fact that the lines in Fig. 9.3 are so nearly straight is very striking.

In Fig. 9.3 the line labeled N_α appears to terminate with the second point. On the other two, labeled Δ_δ and N_γ, there may be more resonances at higher masses which have not yet been observed. Recall that at the higher masses the resonances are masked by a large non-resonant background. The higher the mass, the harder they are to observe. It is possible that these two lines continue on to much higher masses. They may then terminate or turn over (as predicted by one theory to be discussed in the next section). An interesting speculation is that they continue indefinitely. This would mean an infinite number of resonances! As each angular momentum state comes within the range of the π-N force it is good for one resonance. Higher energies and more precise experiments will be needed to distinguish between the various possibilities.

9.3 Recent Theoretical Ideas

Regge Poles. The straight lines of Fig. 9.3 suggest the possibility that the scattering amplitude may be an analytic function of the total angular momentum ℓ considered to be a continuous (perhaps complex) variable [Ref. 4]. Consider the scattering of two spinless particles of non-relativistic energies where the Schroedinger equation is applicable. The wave equation can be solved for arbitrary complex angular momentum ℓ. The solutions at noninteger ℓ are no longer physical states for the usual reason that the angular part of the wave function exhibits singularities. Nevertheless, the partial wave amplitude a (ℓ, k) can be determined at arbitrary complex ℓ by means of the radial wave equation,

$$-\frac{d^2\psi}{dr^2} + \left[\frac{\ell(\ell+1)}{r^2} + V(r) - E\right]\psi = 0 \tag{9.11}$$

The scattering amplitude is given by eq. (2.5),

$$F(\theta, k) = \frac{1}{k}\sum_{\ell=0}^{\infty}(2\ell+1)\,a(\ell, k)\,P_\ell(\cos\theta) \tag{2.5}$$

where $a(\ell, k)$ has been defined by eq. (2.6),

$$a(\ell, k) = \frac{e^{2i\delta(\ell, k)} - 1}{2i} \tag{2.6}$$

The index ℓ is now written as a variable. The solutions of the angular equation can be extended to complex ℓ and complex z by using Schläfli's representation for the Legendre functions,

$$P_\ell(z) = \frac{1}{2\pi i 2^\ell}\int_c \frac{(x^2-1)^\ell}{(x-z)^{\ell+1}}\,dx \tag{9.12}$$

where the contour C is as shown in Fig. 9.4. It is easy to show that expression (9.12) satisfies the Legendre equation, [Ref. 5]

$$\frac{d}{dz}\left[(1-z^2)\frac{dP_\ell(z)}{dz}\right] + \ell(\ell+1)P_\ell(z) = 0 \tag{9.13}$$

Thus, all the quantities on the right hand side of eq. (2.5) can be interpreted to be continuous functions of ℓ. Now $F(\theta, k)$ can be written as a contour integral,

$$F(\theta, k) = \frac{i}{2}\,\frac{1}{k}\int_C (2\ell+1)\,a(\ell, k)\,\frac{P_\ell(-\cos\theta)}{\sin\pi\ell}\,d\ell \tag{9.14}$$

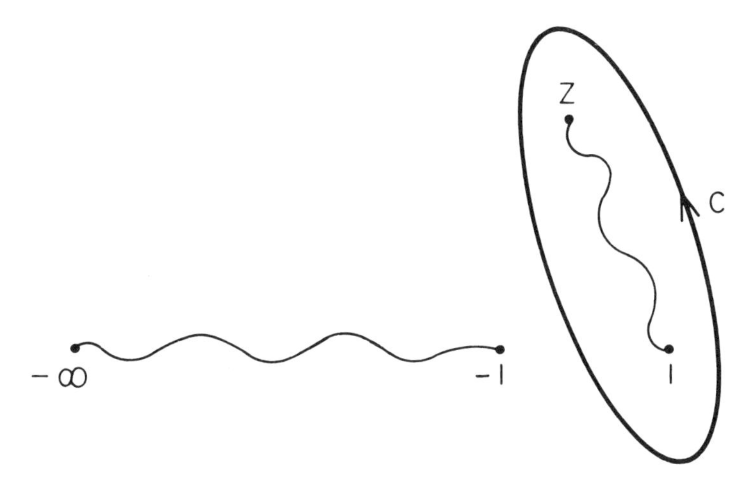

FIG. 9.4 Contour used in Schläfli's representation of the Legendre functions.

where the contour C must include all positive integral values of ℓ as shown in Fig. 9.5. Formula (9.14) is called the Sommerfeld-Watson transformation. In eq. (9.14) ℓ is a continuous complex variable. The integrand has poles at integer ℓ due to the factor $\sin \pi\ell$ in the denominator. Each term in the partial wave summation, given by eq. (2.5), is the residue of a pole in the contour

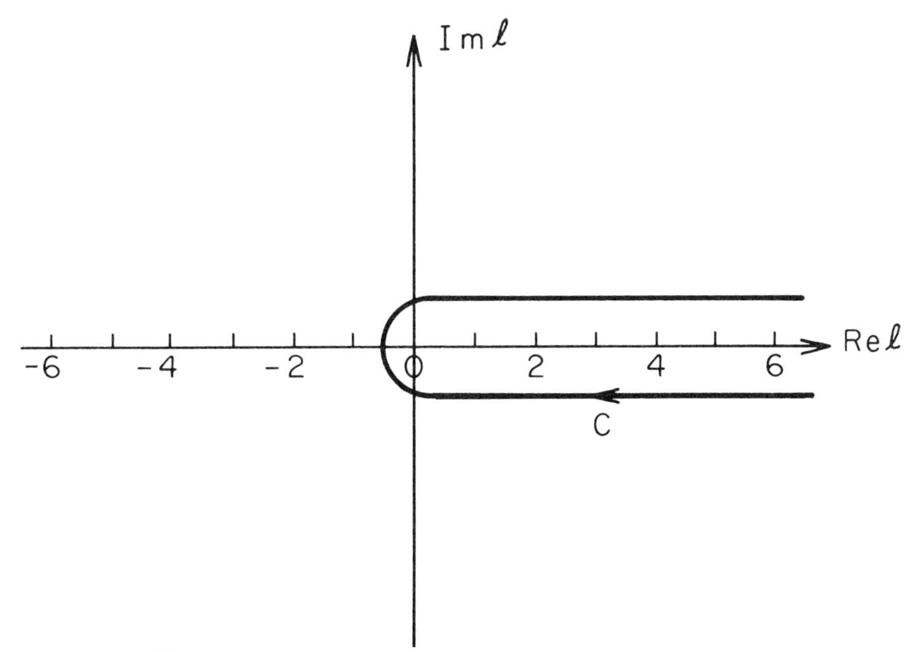

FIG. 9.5 Contour used in the Sommerfeld-Watson transformation.

integration. Near a pole of the integrand at $\ell = n \geqslant 0$, n an integer, we have,

$$\frac{P_\ell(-\cos\theta)}{\sin \pi\ell} \cong \frac{P_n(-\cos\theta)}{\pi(\ell - n)(-1)^n} = \frac{P_n(\cos\theta)}{\pi(\ell - n)} \tag{9.15}$$

which gives the correct residue.

We would now like to expand the contour of Fig. 9.5 in order to exhibit explicitly the other possible singularities of the scattering amplitude $a(\ell, k)$. Using the contour of Fig. 9.6 Regge was able to show [Ref. 6]:

1) that under very general conditions the poles of $a(\ell, k)$ at $\mathrm{Re}\,\ell > -\frac{1}{2}$ lie on the real axis when below threshold and above the real axis when above

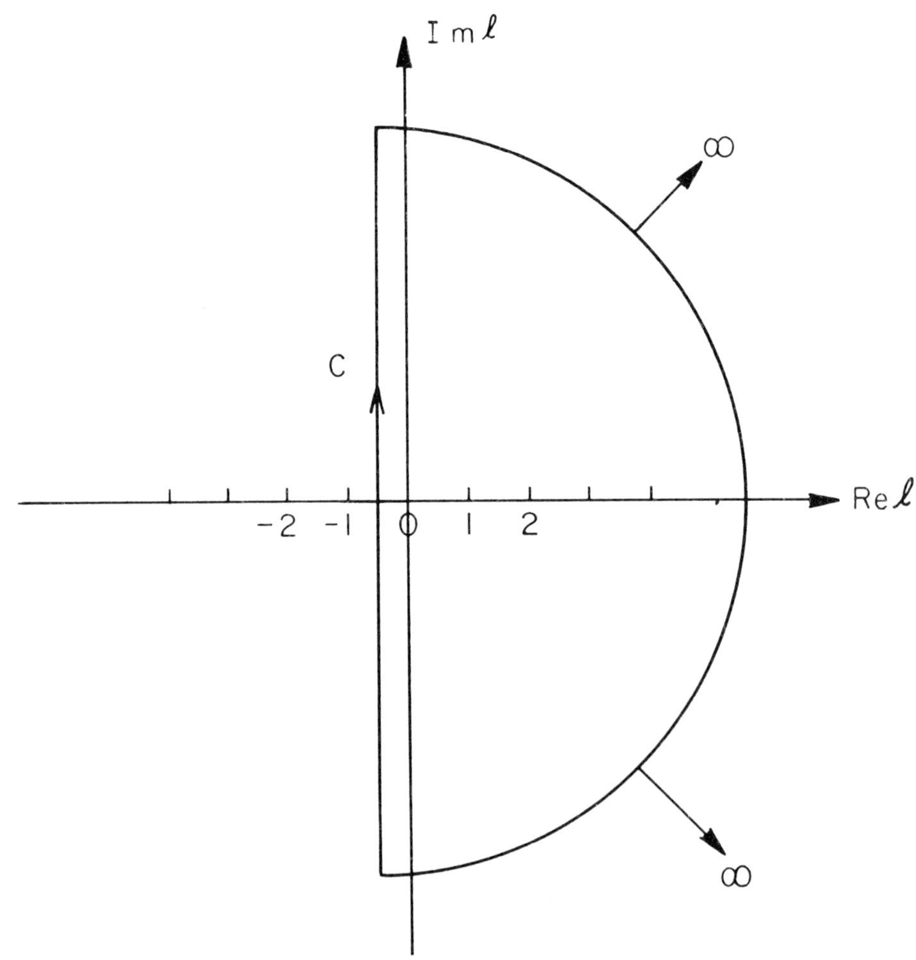

FIG. 9.6 The expanded contour used by Regge in the Sommerfeld-Watson transformation.

threshold,

2) for potentials which can be written as a sum (or integral over) Yukawa potentials $a(\ell, k)$ has no other singularities except a finite number of poles for $\mathrm{Re}\,\ell > -\frac{1}{2}$,
3) for sums of Yukawa potentials the contribution of the semicircle at infinity vanishes in the integral of eq. (9.14). (It does not vanish for a square well, for example.)

Thus, eq. (9.14) can be re-written,

$$F(\theta,k) = \frac{i}{2k} \int_{-1/2-i\infty}^{-1/2+i\infty} (2\ell+1)\, a(\ell,k) \frac{P_\ell(-\cos\theta)}{\sin\pi\ell}\, d\ell + \sum_i \frac{\beta_i(k)\, P_{\alpha_i}(k)\,(-\cos\theta)}{\sin\pi\alpha_i(k)} \tag{9.16}$$

where $\beta_i(k)$ stands for the integral $(2\ell+1)\, a(\ell,k)\, d\ell$ around the pole at complex angular momentum α_i. The special significance of $\mathrm{Re}\,\ell = -\frac{1}{2}$ results from the fact that at this value of ℓ, the irregular solution of eq. (9.11),

$$\lim_{r\to 0} \psi_I \propto r^{-\ell} \tag{9.17}$$

crosses the regular solution,

$$\lim_{r\to 0} \psi_R \propto r^{\ell+1} \tag{9.18}$$

The poles of $a(\ell, k)$ giving rise to the Σ_i in eq. (9.16) are called Regge poles. These poles are functions of ℓ and k. Their movement describes the so called Regge trajectories. The curves of Fig. 9.3 are examples of Regge trajectories.

To project out a single partial wave from the scattering amplitude we write,

$$F_n(k) = \int_{-1}^{1} P_n(\cos\theta)\, F(\theta, k)\, d\cos\theta \tag{9.19}$$

where n is a real integer. If eqs. (2.5) and (9.16) are substituted into eq. (9.19), the result is,

$$F_n(k) = \frac{1}{k}\, a_n(k) + 2 \sum_i \frac{\beta_i(k)}{\pi[\alpha_i(k)-n]\;[\alpha_i(k)+n+1]} \tag{9.20}$$

Use was made of the identity,

$$\int_{-1}^{1} P_n(\cos\theta)\, P_\alpha(-\cos\theta)\, d\cos\theta = \frac{\sin\pi\alpha}{\pi(\alpha-n)\,(\alpha+n+1)} \tag{9.21}$$

Assume that for some energy the real part of one of the α_i is equal to n. That is,

$$\mathrm{Re}\;\alpha_j(\omega_n) = n \tag{9.22}$$

where ω_n is the total energy. In the region near ω_n expand α_j in a Taylor series,

$$\alpha_j(\omega) = n + (\omega - \omega_n) \frac{d}{d\omega} \operatorname{Re} \alpha_j \Big|_{\omega_n} + i \operatorname{Im} \alpha_j(\omega_n) \tag{9.23}$$

Substitution into the jth term of the Σ_i in eq. (9.20) gives near $\omega = \omega_n$,

$$\frac{2\beta_j(\omega_n)}{\pi[\alpha_j(\omega_n) + n + 1] \dfrac{d}{d\omega} \operatorname{Re} \alpha_j \Big|_{\omega_n} \left(\omega - \omega_n + i \dfrac{\Gamma}{2}\right)} \tag{9.24}$$

where,

$$\frac{\Gamma}{2} = \frac{\operatorname{Im} \alpha_j(\omega_n)}{\dfrac{d}{d\omega} \operatorname{Re} \alpha_j \Big|_{\omega_n}} \tag{9.25}$$

Eq. (9.24) is just the form of a Breit-Wigner resonance with width Γ given by eq. (9.25).

Below threshold the poles lie on the real axis in the ℓ-plane. By eq. (9.25) $\Gamma = 0$. As discussed in Sect. 4.1 this corresponds to a bound state as it should.

It is *assumed* that the major results of the Regge idea can be carried over into non-relativistic scattering where the non-relativistic concept of a potential is no longer applicable. For the π-N state Regge poles occur at half integral values of Rej where j is now the *total* angular momentum. On each Regge trajectory there are poles at *every other* half integer as can be seen from Fig. 9.3. This is because exchange forces require that the poles on each trajectory have the same parity.

While the Regge theory gives a natural expression to the smooth curves of Fig. 9.3, it does not explain its two most notable features, namely, (a) why the trajectories should be so straight and (b) why the different trajectories should be so nearly parallel. The non-relativistic Regge theory makes the highly interesting prediction that the trajectories of Fig. 9.3 should eventually turn over and return to Re$j = \frac{1}{2}$. It does not predict at what energy this should begin to occur. However, in the extrapolation to relativistic energies where more and more inelastic states come into play, this prediction may break down. The search for higher resonances is thus an extremely important problem for experimentalists to pursue.

SU_3

It is a remarkable fact that the eight familiar baryons n, p, Λ°, Σ^{+}, Σ°, Σ^{-}, Ξ^{-}, Ξ° all have the same spin and parity. This suggests that they are members of a single super-multiplet. This is assumed to result from a higher symmetry group of elementary particles. The mass differences are presumed to result from an interaction which breaks this higher symmetry.

The isotopic spin group is SU_2, the group of all unitary 2×2 matrices with unit determinant. The higher symmetry group proposed independently by M. Gell-Mann and Y. Ne'eman is SU_3, the group of all unitary 3×3 matrices with unit determinant [Ref. 7].

A unitary transformation can be written

$$U = e^{iA} \tag{9.26}$$

where A is Hermitian. Furthermore det $U = 1$ implies Tr $A = 0$. There are three independent 2×2 Hermitian traceless matrices. There are thus three components of isotopic spin. There are eight independent 3×3 Hermitian traceless matrices. There are thus eight components of "unitary spin." Unitary symmetry corresponds to the unitary group in three dimensions in the same way that charge independence corresponds to the unitary group in two dimensions. Unitary symmetry is the simplest generalization of charge independence.

A typical set of unitary spin matrices is,[1]

$$A_1 = \begin{pmatrix} 0 & 1 & 0 \\ 1 & 0 & 0 \\ 0 & 0 & 0 \end{pmatrix}, \quad A_2 = \begin{pmatrix} 0 & -i & 0 \\ i & 0 & 0 \\ 0 & 0 & 0 \end{pmatrix}, \quad A_3 = \begin{pmatrix} 1 & 0 & 0 \\ 0 & -1 & 0 \\ 0 & 0 & 0 \end{pmatrix},$$

$$A_4 = \begin{pmatrix} 0 & 0 & 1 \\ 0 & 0 & 0 \\ 1 & 0 & 0 \end{pmatrix}, \quad A_5 = \begin{pmatrix} 0 & 0 & -i \\ 0 & 0 & 0 \\ i & 0 & 0 \end{pmatrix}, \quad A_6 = \begin{pmatrix} 0 & 0 & 0 \\ 0 & 0 & 1 \\ 0 & 1 & 0 \end{pmatrix},$$

$$A_7 = \begin{pmatrix} 0 & 0 & 0 \\ 0 & 0 & -i \\ 0 & i & 0 \end{pmatrix}, \quad A_8 = \begin{pmatrix} \frac{1}{\sqrt{3}} & 0 & 0 \\ 0 & \frac{1}{\sqrt{3}} & 0 \\ 0 & 0 & \frac{-2}{\sqrt{3}} \end{pmatrix} \tag{9.27}$$

It can be shown that at most two of the matrices can be diagonal at the same time. These have been chosen to be A_3 and A_8. These matrices operate on a fundamental triplet of particles (q_1, q_2, q_3) named "quarks" by Gell-Mann. It is assumed that a meson is a bound state of a quark-antiquark pair $q\bar{q}$, and that a baryon is a bound state of three quarks qqq.

Since the eight familiar baryons have isotopic spins 0, $\frac{1}{2}$, and 1, it is clear that two of the quarks must have isotopic spin $\frac{1}{2}$ and the third isotopic spin 0. If q_1 and q_2 are the isotopic spin doublet then A_1, A_2 and A_3 are just the isotopic spin operators. Further, one quark must have strangeness -1. A

[1]These unitary spin operators also obey the right commutation rules to form the eight generators of the Lie algebra of SU_3.

little thought shows that in order to obtain the correct quantum numbers out of $\bar{q}q$ for the mesons and qqq for the baryons q_3 must have strangeness 1 and q_1 and q_2 strangeness 0. It is clear also that the three quarks must have baryon number $\frac{1}{3}$ in order to add up to 1 for the various baryons. With these assignments of strangeness and baryon number, we see that A_8 is proportional to the hypercharge,

$$Y = N + S = \frac{1}{\sqrt{3}} A_8 \tag{9.28}$$

where Y = hypercharge, N = baryon number, and S = strangeness. The new operators $A_4, \ldots, A_7$ obey the rules $|\Delta\vec{T}| = \frac{1}{2}$ and $|\Delta Y| = 1$. Under SU_3 symmetry the strong interactions, besides conserving the three components of isotopic spin and the hypercharge, also approximately conserve four more operators $A_4, \ldots A_7$.

We mention one final property of the quarks. Using the Gell-Mann-Nishijima formula $Q = I_3 + \frac{1}{2} Y$ it is seen that the quarks have charge $(\frac{2}{3}, -\frac{1}{3}, -\frac{1}{3})$ respectively. Although experimentalists have searched diligently for fractionally charged particles, none have been found. This is somewhat of an embarrassment for the theory.

The quarks transform under SU_3 according to an irreducible 3-dimensional representation **3**. The antiquarks transform according to a second 3-dimensional representation $\bar{\mathbf{3}}$ (not equivalent to **3**). Mesons and baryons will transform according to products of these representations $\bar{\mathbf{3}} \times \mathbf{3}$ and $\mathbf{3} \times \mathbf{3} \times \mathbf{3}$ respectively. These products will in general contain more than one irreducible representation. The results can be found in standard group multiplication tables [Ref. 8]. For SU_3 we have,

$$\begin{aligned} &\text{Mesons} \quad \bar{\mathbf{3}} \times \mathbf{3} = \mathbf{8} + \mathbf{1} \\ &\text{Baryons} \quad \mathbf{3} \times \mathbf{3} \times \mathbf{3} = \mathbf{10} + \mathbf{8} + \mathbf{8} + \mathbf{1} \end{aligned} \tag{9.29}$$

Eqs. (9.29) express the most spectacular results of SU_3 theory. These equations predict that the supermultiplets of mesons are octets and singlets only and the supermultiplets of baryons are decuplets, octets and singlets. These supermultiplets are easily identified since the particles within each supermultiplet will have the same spin and parity. *The predictions of eqs. (9.29) appear to agree with experiment.*

The octet of familiar baryons has already been mentioned. There is in addition a decuplet containing $N^*(1238)^{++}, N^*(1238)^{+}, N^*(1238)^{\circ}, N^*(1238)^{-}$, $Y_1^*(1385)^{+}$, $Y_1^*(1385)^{\circ}$, $Y_1^*(1385)^{-}$, $\Xi_{1/2}^*(1530)^{\circ}$, $\Xi_{1/2}^*(1530)^{-}$, $\Omega^{-}(1674)$. There are two known meson octets, one containing the familiar $\pi^+, \pi^-, \pi^\circ, K^+$, K°, $\bar{K}^\circ$, K^-, η and a vector meson octet. The $\Lambda(1405)$ appears to be a singlet as does the $\eta'(959)$. The ϕ and ω mesons appear to be linear combinations of a vector singlet and the $T = Y = 0$ member of the vector octet. This must result from a symmetry breaking interaction. Supermultiplets with multiplicities other than those predicted by eqs. (9.29) have not been observed.

If the Regge pole idea and SU_3 theory are both valid, then all the members of a given representation must lie on Regge trajectories having poles at the same values of the spin. Thus, the N_α trajectory of Fig. 9.3 implies the existence of a $j = \frac{5}{2}$ even parity octet. In a similar way there must be decuplets of particles with spins $j = \frac{7}{2}$, $\frac{11}{2}$, and $\frac{15}{2}$ all with even parity. Experimentally there are resonances in Λ-π, Σ-π and Ξ-π systems which fit into some of these supermultiplets [Ref. 9]. However, none of the supermultiplets are complete. For example the N^* (1688), Y_0^* (1820) and a possible Y_1^* (1910) make up six members of an octet. There are $\Xi_{1/2}^*$ resonances at 1815 and 1930 MeV with spins as yet unknown. If one turns out to be $\frac{5}{2}^+$ then both octets on the N_α trajectory will be filled. The N_γ trajectory is probably an octet. There are six $\frac{3}{2}^-$ states known: N^* (1525) (2 states), Y_0^* (1520) (1 state) and Y_1^* (1660) (3 states). The spin of the last is not yet definite. Again one of the two $\Xi_{1/2}^*$ states with unknown spin and parity may turn out to be $\frac{3}{2}^-$. Then this octet will be filled. The multiplets corresponding to the other states on the trajectories of Fig. 9.3 have few experimentally identified entries. The search for resonances with spins and parities needed to fill up these multiplets remains an important problem.

Bootstraps. In Sect. 9.1 it was pointed out that the force responsible for the $T = \frac{3}{2}$, $j = \frac{3}{2}$ resonance at $T_\pi = 195$ MeV is due primarily to the exchange of a nucleon. That is the N^* (1238) is approximately a pion-nucleon composite held together primarily by exchange of a nucleon. Chew has suggested the intriguing possibility that the converse may be true [Ref. 10]. A nucleon may be approximately a composite of a pion and a nucleon bound together to a large extent by exchange of an N^* (1238). Calculation shows that this force should be strongly attractive. This idea has been generalized to include the other resonances as well as using results from Regge pole and SU_3 theory [Ref. 11]. In order that these particle states generate each other, their masses and coupling constants must be *mutually consistent.* It is possible that this requirement of consistency will demand unique masses and coupling constants. These can then all be calculated from some single mass such as the pion mass and a single coupling constant. Such a consequence would be of enormous significance if true. The detailed calculations are very complex. So far the results have not been impressive. This may be due to the severe approximations needed to obtain a numerical result from the calculations.

Summary. The three ideas discussed above are all extremely interesting. All of the logical consequences of each of them have not yet been extracted. Each has its difficulties. It is hard to reconcile Regge pole theory with certain features of high energy cross sections. In SU_3 theory all the particles in a given multiplet should have similar mass. The differences would be due presumably to some symmetry breaking interaction. If this interaction is large then grouping the particles into multiplets would not be meaningful. The mass differences among the various members of a multiplet, especially meson multiplets, are in fact sometimes quite large. On the other hand grouping particle states into supermultiplets corresponding to irreducible representations of SU_3 appears to be correct. Why this is so is somewhat of a mystery. The bootstrap calculations show only qualitative agreement with the observed spectrum of masses.

It is to be hoped that these various problems will be clarified in the future. Out of the ideas outlined above a comprehensive theory may emerge. In the meantime experimental verification of the simple predictions mentioned above is extremely important.

REFERENCES

1. G. F. Chew and F. E. Low, *Physical Review 101,* 1570 (1956).
2. See for example, *Elementary Particle Physics* by Gunnar Källén, Addison-Wesley (1964) or *An Introduction to Relativistic Quantum Field Theory,* by Silvan Schweber Harper & Row (1961) or *The Theory of Elementary Particles,* by J. Hamilton, Oxford (1959).
3. F. E. Low, *Physical Review 97,* 1392 (1955).
4. For an introduction to Regge pole theory see *Regge Poles and S-Matrix Theory,* by S. C. Frautschi, W. A. Benjamin, Inc. (1963).
5. E. T. Whittaker and G. N. Watson, *A Course in Modern Analysis,* Cambridge University Press, Ch. XV (1952).
6. T. Regge, *Nuovo Cimento 14,* 951 (1959).
7. For an introduction to this theory see, *The Eightfold Way,* by M. Gell-Mann and Y. Ne'eman, W. A. Benjamin, Inc. (1964).
8. R. E. Behrends, J. Dreitlein, C. Fronsdal, and W. Lee, *Rev. Mod. Phys. 34,* 1 (1962).
9. For a summary of all the experimental data on resonant states see, A. H. Rosenfeld, N. Barash-Schmidt, A. Barbaro-Galtieri, L. R. Price, P. Söding, C. G. Wohl, M. Roos and W. J. Willis, *Rev. Mod. Phys. 40,* 77 (1968). The notation used in this chapter for resonant states is the same as in this reference.
10. G. F. Chew, *Phys. Rev. Letters 9,* 233 (1962).
11. P. Carruthers, *Phys Rev. Letters 12,* 259 (1964).

APPENDIX A

We expand,

$$\frac{d\sigma}{d\Omega} = \lambda\!\!\!^{-2} \sum_n A_n f_n(\cos\theta) \qquad (A.1)$$

$$\frac{d\sigma}{d\Omega}\vec{P} = \vec{n}_\perp \lambda\!\!\!^{-2} \sum_n B_n g_n(\cos\theta) \qquad (A.2)$$

for Case 1 of Sect. 5.1 ($P_i = 0$) where the f_n and g_n are a complete set of functions and

$$\vec{n}_\perp = \frac{\vec{k}_{\text{out}} \times \vec{k}_{\text{in}}}{|\vec{k}_{\text{out}} \times \vec{k}_{\text{in}}|} \qquad (A.3)$$

Further,

$$A_n = \sum_{\ell,j\ \ell',j'} \alpha_n(\ell,j,\ell',j')\,\text{Re}(a_{\ell,j}^{*}\,a_{\ell',j'}) \qquad (A.4)$$

$$B_n = \sum_{\ell,j\ \ell',j'} \beta_n(\ell,j,\ell',j')\,\text{Im}(a_{\ell,j}^{*}\,a_{\ell',j'}) \qquad (A.5)$$

where the $a_{\ell j}$ are the partial wave amplitudes defined in eq. (2.44). The following tables list the coefficients α_n and β_n for all partial waves through $G_{7/2}$. The functions f_n and g_n are:

$$f_n(\cos\theta) = P_n(\cos\theta) \qquad \text{Table A.1}$$

$$g_n(\cos\theta) = P_n^1(\cos\theta) = \sin\theta\,\frac{d}{d\cos\theta}\,P_n(\cos\theta) \qquad \text{Table A.2}$$

$$f_n(\cos\theta) = \cos^n\theta \qquad \text{Table A.3}$$

$$g_n(\cos\theta) = \sin\theta\,\cos^n\theta \qquad \text{Table A.4}$$

The tables show the coefficients for linear combinations of partial waves which exhibit explicitly the Minami ambiguity. For more extensive tables see L. E. Olson and W. P. Trower, *Journ. Nat. Sci. and Math.* 6, (1967). (The coefficients for $j = j'$ and $\ell' = \ell + 1$ in those tables should be divided by 2).

TABLE A.1

Coefficients α_n in a Legendre Polynomial Expansion

$\ell, j, \ell', j' \quad n \longrightarrow$	0	1	2	3	4	5	6	7
S1S1 + P1P1	1							
S1P1		2						
S1P3 + P1D3		4						
S1D3 + P1P3			4					
S1D5 + P1F5			6					
S1F5 + P1D5				6				
S1F7 + P1G7				8				
S1G7 + P1F7					8			
P3P3 + D3D3	2		2					
P3D3		4/5		36/5				
P3D5 + D3F5		36/5		24/5				
P3F5 + D3D5			12/7		72/7			
P3F7 + D3G7			72/7		40/7			
P3G7 + D3F7				8/3		40/3		
D5D5 + F5F5	3		24/7		18/7			
D5F5		18/35		16/5		100/7		
D5F7 + F5G7		72/7		8		40/7		
D5G7 + F5F7			8/7		360/77		200/11	
F7F7 + G7G7	4		100/21		324/77		100/33	
F7G7		8/21		24/11		600/91		9800/429

TABLE A.2

Coefficients β_n in a Legendre Polynomial Expansion

$\ell, j, \ell', j' \quad n \longrightarrow$	1	2	3	4	5	6	7
S1P1	2						
S1P3 − P1D3	−2						
S1D3 − P1P3		2					
S1D5 − P1F5		−2					
S1F5 − P1D5			2				
S1F7 − P1G7			−2				
S1G7 − P1F7				2			
P3D3	8/5		12/5				
P3D5 − D3F5	−18/5		−2/5				
P3F5 − D3D5		10/7		18/7			
P3F7 − D3G7		−24/7		−4/7			
P3G7 − D3F7			4/3		8/3		
D5F5	54/35		8/5		20/7		
D5F7 − F5G7	−36/7		−2/3		−4/21		
D5G7 − F5F7		4/3		18/11		100/33	
F7G7	32/21		16/11		160/91		1400/429

TABLE A.3

Coefficients α_n in a $\cos\theta$ Series Expansion

ℓ, j, ℓ', j' $n \longrightarrow$	0	1	2	3	4	5	6	7
S1S1 + P1P1	1							
S1P1		2						
S1P3 + P1D3		4						
S1D3 + P1P3	−2		6					
S1D5 + P1F5	−3		9					
S1F5 + P1D5		−9		15				
S1F7 + P1G7		−12		20				
S1G7 + P1F7	3		−30		35			
P3P3 + D3D3	1		3					
P3D3		−10		18				
P3D5 + D3F5				12				
P3F5 + D3D5	3		−36		45			
P3F7 + D3G7	−3		−6		25			
P3G7 + D3F7		21		−110		105		
D5D5 + F5F5	9/4		−9/2		45/4			
D5F5		45/2		−117		225/2		
D5F7 + F5G7		9		−30		45		
D5G7 + F5F7	−9/2		207/2		−675/2		525/2	
F7F7 + G7G7	9/4		45/4		−165/4		175/4	
F7G7		−81/2		795/2		−815		1225/2

TABLE A.4

Coefficients β_n in a $\cos\theta$ Series Expansion

ℓ, j, ℓ', j' $n \longrightarrow$	0	1	2	3	4	5	6
S1P1	−2						
S1P3 − P1D3	2						
S1D3 − P1P3		−6					
S1D5 − P1F5		6					
S1F5 − P1D5	3		−15				
S1F7 − P1G7	−3		15				
S1G7 − P1F7		15		−35			
P3D3	2		−18				
P3D5 − D3F5	3		3				
P3F5 − D3D5		15		−45			
P3F7 − D3G7		6		10			
P3G7 − D3F7	−3		60		−105		
D5F5	−9		63		−225/2		
D5F7 − F5G7	9				15/2		
D5G7 − F5F7		−63/2		210		−525/2	
F7G7	9		−315/2		1275/2		−525/2

APPENDIX B

Table B.1 displays the functions $f_\alpha(x)$, $f_{\alpha\beta}(x)$ and $g_{\alpha\beta}(x)$ defined in euqations 8.3 and 8.4. Multipoles thru $f_{5/2}$ are shown. The entries with $\alpha < \beta$ are $f_{\alpha\beta}(x)$ and those with $\alpha > \beta$ are $g_{\alpha\beta}(x)$. Those with $\alpha = \beta$ are $f_\alpha(x)$.

TABLE B.1

Pion Photoproduction Angular Distributions ($\alpha \leqslant \beta$) and Polarizations ($\alpha > \beta$).†

Final state	α ↓ β →	N_α ↓ N_β →	E_1^-	M_1^-	M_1^+	E_2^-	E_1^+	M_2^-	M_2^+	E_3^-	E_2^+	M_3^-
			1	1	1	$\sqrt{3}$	1	$\sqrt{3}$	$\sqrt{3}$	$\sqrt{\frac{3}{8}}$	$\sqrt{3}$	$\sqrt{\frac{3}{8}}$
$s\frac{1}{2}$	E_1^-	1	1	$-2x$	$2x$	$2x$	$3x^2-1$	$1-3x^2$	$-(1-3x^2)$	$-4(1-3x^2)$	$-x(3-5x^2)$	$4x(3-5x^2)$
$p\frac{1}{2}$	M_1^-	1	2	1	$1-3x^2$	$(1-3x^2)$	$-2x$	$2x$	$x(3-5x^2)$	$4x(3-5x^2)$	$(1-3x^2)$	$-4(1-3x^2)$
$p\frac{3}{2}$	M_1^+	1	1	$-3x$	$\frac{1}{2}(5-3x^2)$	$-(1-3x^2)$	$+2x$	$2x(2-3x^2)$	$2x(3-2x^2)$	$-4x(3-5x^2)$	$-(1-3x^2)$	$-5+42x^2-45x^4$
$p\frac{3}{2}$	E_2^-	$\sqrt{3}$	1	$-3x$	0	$\frac{1}{2}(1+x^2)$	$-2x(2-3x^2)$	$2x(1-2x^2)$	$-2x(1-2x^2)$	$4x(1+x^2)$	$1-9x^2+10x^4$	$-(1-18x^2+25x^4)$
$d\frac{3}{2}$	E_1^+	1	$-3x$	1	-4	$2(1-3x^2)$	$\frac{1}{2}(5-3x^2)$	$(1-3x^2)$	$-(1-3x^2)$	$5-42x^2+45x^4$	$2x(3-2x^2)$	$4x(3-5x^2)$
$d\frac{3}{2}$	M_2^-	$\sqrt{3}$	$3x$	-1	$-2(1-3x^2)$	$4x^2$	0	$\frac{1}{2}(1+x^2)$	$-1+9x^2-10x^4$	$-(1-18x^2+25x^4)$	$2x(1-2x^2)$	$4x(1+x^2)$
$d\frac{5}{2}$	M_2^+	$\sqrt{3}$	$2x$	$1-5x^2$	$2-x^2$	x^2	$5x$	$5x(1-2x^2)$	$\frac{1}{2}(1+6x^2-5x^4)$	$(1-18x^2+25x^4)$	$-2x(1-2x^2)$	$-19x+86x^3-75x^5$
$d\frac{5}{2}$	E_3^-	$\sqrt{\frac{3}{8}}$	$8x$	$4(1-5x^2)$	$-(1-5x^2)$	$3+x^2$	$5x(9x^2-5)$	$5x(1-5x^2)$	0	$5+6x^2+5x^4$	$19x-86x^3+75x^5$	$-2x(1-5x^2)(11-15x^2)$
$f\frac{5}{2}$	E_2^+	$\sqrt{3}$	$1-5x^2$	$2x$	$-5x$	$5x(1-2x^2)$	x^2-2	x^2	$-6x^2$	$-3(1-18x^2+25x^4)$	$\frac{1}{2}(1+6x^2-5x^4)$	$-(1-18x^2+25x^4)$
$f\frac{5}{2}$	M_3^-	$\sqrt{\frac{3}{8}}$	$4(5x^2-1)$	$-8x$	$-5x(5-9x^2)$	$-5x(1-5x^2)$	$-(1-5x^2)$	$-(3+x^2)$	$3(1-18x+25x^2)$	$6(1-5x^2)^2$	0	$5+6x^2+5x^4$

†The $\alpha\beta$ element must be multiplied by the normalization factor $N_\alpha N_\beta$.

INDEX